ZOOLOGY
Laboratory Workbook

Ninth Edition

Barbara J. Schumacher, M.S.
San Jacinto College Central

Marlin O. Cherry, Ed.D.
Vice Chancellor of Instruction—Retired
San Jacinto College District

KENDALL/HUNT PUBLISHING COMPANY
4050 Westmark Drive   Dubuque, Iowa 52002

Cover image courtesy Corel.

Contents

Preface

This laboratory manual is designed as a one-semester course in zoology and to accompany any recently published general biology or zoology textbook. Various exercises cover most of the biological disciplines such as cytology, morphology, taxonomy, anatomy and physiology. The workbook focuses on the traditional general survey of the animal kingdom.

Unit I introduces the student to the molecular structure of the chemical building blocks of life. Since the microscope is integral to observation in the biology laboratory, proper use of the microscope and the production of wet mount slides are stressed. Basic principles of microscopy, cytology, and histology are addressed in this first section through the observation of various tissue slide preparations.

Unit II investigates the processes of cell division, both mitosis and meiosis. Following the production of sex cells, focus is placed on the early embryological development of typical vertebrates.

Unit III includes the survey of animal-like protists and the invertebrate animals. Important structural characteristics and typical representatives of each phylum will be studied with an emphasis on the levels of organization. Although preserved and bottled samples are available, live specimens should be viewed whenever possible.

Unit IV addresses many of the familiar vertebrate animals with emphasis on the common vertebrate representative, the frog. A major portion of this unit is devoted to its study. Although the amphibian anatomy and physiology are detailed, comparison and references are also made to human structure and function.

Throughout the manual there are numerous photographs and drawings of animals and/or their structures; however, they are not intended to replace the study or observation of real animals. Since drawing is one of the oldest methods of recording, students should attempt to sketch their own pictures.

The workbook is written so that the student may proceed with the laboratory work with just a short introduction by the instructor. The student will find it possible to complete the daily assignment with the aid of written directions, charts, textbook, and other visible means. The instructor is available to aid students having difficulty and to clarify important terminology or procedures.

Barbara J. Schumacher, M.S.
Marlin O. Cherry, Ed.D.
Pasadena, Texas
March, 2003

Acknowledgments

We would like to thank Patricia Steinke, Larry Petersen, Dan Penney and Robyn Courville for proofreading the manual and providing many helpful suggestions and Robert Kirkley for producing the chemical structures shown in Figure 2.1. Black and white photographs are new productions by the authors or from the collections of both Marlin O. Cherry and the late Billy J. Hart.

General Instructions

This workbook is designed to present the student with material that is discussed in the lecture portion of this course. The student should not consider the laboratory as a separate course, but a period to gain a more intimate knowledge of lecture topics. It is important that students prepare for each exercise by reading the material before coming to the lab. Each student is expected to complete all required work. Questions are given throughout the various exercises to emphasize certain principles and draw attention to particular structures or activities.

At the beginning of the laboratory period each student will be given a permanent table, seat and group number. The microscope, slides and other equipment which will be used also bears a corresponding number to the assigned group. A laboratory fee is charged for this course, but students will be asked to pay for breakage under certain conditions. Report damaged items to the instructor at the beginning of the laboratory and the person(s) using them in the previous section will be charged. If damage or breakage occurs during the working period, the individual responsible will be notified of the repair or replacement cost.

In order to avoid confusion, try to be in the laboratory on time. Latecomers will miss important instructions needed for the laboratory work. If a student desires to spend more review time in the lab, check with the instructor for the hours set aside for this purpose. Each student is expected to complete the assigned work within the lab period time frame.

LABORATORY RULES

1. Polite and attentive behavior is expected.

2. No smoking, food or drinks are allowed in the lab room.

3. At the end of the period, clean and return all equipment and material to the proper place. Put all paper debris and other biological waste in the marked containers provided for these purposes.

4. Please push the chair or stool under the table at the end of the period.

5. Laboratories are set up for cooperation by partners. Each should share in the microscope viewing, preparation of wet mounts, clean-up, etc.

6. Absolutely NO equipment is to be taken from the laboratory. Anyone found removing material will be reported to the proper authorities.

7. Slides, charts, jars and live specimens are also presented for observation. Students are responsible for viewing these materials during the period.

8. Exercises are assigned each lab period; therefore, bring this manual to every class.

9. Supervised lab time will be made available for students who feel the need for additional review.

10. No student should "sit-in-on" or otherwise interrupt the scheduled lab time of another instructor or section.

11. Since the laboratory directions are written for the benefit of the students, it is worthwhile to follow them closely. Each student should develop a certain independence in completing the assigned work. However, the instructor is anxious and ready to help, so call on him/her freely.

EQUIPMENT AND MATERIALS

In addition to your laboratory manual, each working group is expected to have a dissecting kit. Once partners have been established, a decision should be made on the purchase, storage and availability of this equipment.

Since lab dissections are often "messy", lab coats or aprons are suggested but optional. However, gloves are encouraged for protection from various preservatives. Hand lotions are always effective against the drying effect of preservatives and constant hand-washing.

Unit I
Introduction to Cells and Tissues

EXERCISE 1   The Microscope and Its Use

Learning Objectives

- ✔ Identify the parts of the compound microscope and their functions.
- ✔ Use and care for the compound microscope.
- ✔ Identify the parts of the dissecting microscope and their functions.
- ✔ Use and care for the dissecting microscope.
- ✔ Determine the magnification of an image.
- ✔ Determine the size of objects with the compound microscope.
- ✔ Properly prepare a wet mount slide.

Introduction to the Compound Microscope

A microscope is an optical instrument consisting of one lens (simple microscope) or a system of two or more lenses (compound microscope) for making enlarged or magnified images of minute objects. Only low magnifications can be obtained through the use of a single lens. The **compound** microscope is the type generally used by students for viewing very small objects.

Although many different student compound microscopes are available today (Plate 1.1 A, B, C, D, E), each scope essentially includes: a series of lenses to magnify the object, an illumination source, mechanical parts to support the magnifying and illumination mechanisms and a means to rapidly adjust and focus these parts. It consists of the following main parts:

1. A stand.
2. A body tube with an ocular piece.
3. Rotatable nosepiece containing several objectives.
4. Stage with slide retaining apparatus.
5. Condenser with its associated iris diaphragm.
6. In-base illuminator.
7. Coarse and fine adjustment wheels.

The **body tube** is the upper, tubular portion in which the **ocular** or eyepiece is seated. On many microscopes the body tube is stationary, inclined and includes one eyepiece (monocular) while other models have two eyepieces (binocular) and may swivel around for partner use. The ocular is the topmost optical lens system in a microscope and has a magnifying capability of 10x (times).

At the base of the body tube is a **revolving nosepiece**, a disc-like metal piece which holds the **objectives**. The nosepiece may be equipped with two, three or four **objectives**. If four objectives are present, the shortest one is called the "scanning" power and engraved with a value (generally) of 4X. The next objective is called "low power" and is marked

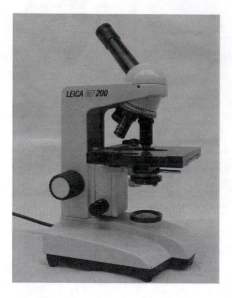

PLATE 1.1 A   Leica

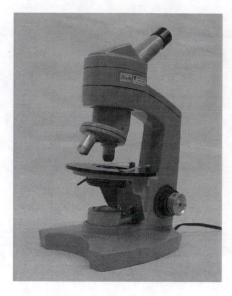

PLATE 1.1 B   American Optical

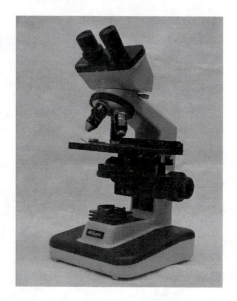

PLATE 1.1 C   Nikon Alphaphot

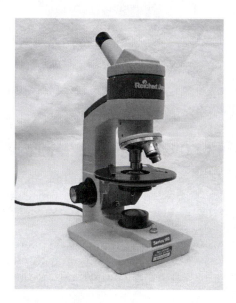

PLATE 1.1 D   Reichert

10X. The "high dry power" objective may be marked 40X or 43X while the longest "high oil immersion" objective is marked 97X or 100X. Thus, the compound microscope has two separate lens systems. The lower objective magnifies and produces an image termed the **real image**. The upper ocular further magnifies the real image into a **virtual image**, approximately equal to the product of the two systems and expressed in terms of "diameter", "power", or "X". Hence, these combined systems give a total magnifying capability of 40X, 100X, 400X (or 430X), 970X (or 1000X), respectively.

The **stage** is a horizontal plate, round or square, with a central opening just above the **illuminator,** or light source. The stage is used to support the object to be observed. Varieties of stage mechanisms exist. Some "floating" stages allow free, easy movement but many "fixed" stages have either: (1) two simple clips or (2) a more elaborate mechanical apparatus with a set of turning screws to hold and position the slide along an X-Y axis.

Below the stage's aperture is a **condenser.** The condenser collects the light from the light source and converges it into a cone of sufficient angle to fill the lens of the objective with light. An **iris diaphragm,** located just beneath the condenser, regulates the size of the aperture in order to control the amount of light striking the specimen.

Turning the two pairs of wheels, **coarse adjustment** and **fine adjustment,** moves the body tube only a short distance at a controlled pace. The larger coarse adjustment knob brings the specimen into general focus while the smaller

PLATE 1.1 E   Nikon E200

fine adjustment knob sharply focuses the image for individual eyesight. Since the focusing knobs on both sides are co-ordinated, the scope is convenient for either right or left handed users. Focusing with either hand leaves the other to write descriptions and observations or draw illustrations.

The part of the microscope that supports the magnifying lens and illuminating portion is known as the **stand.** The bottom of the stand, or **base,** is usually forked and supports all the other parts. Above the base is the **pillar** that connects the stage to the base and the **arm** supports the body tube.

Suggestions for Using the Compound Microscope

1. Use the instrument assigned to you. Report to the instructor without delay any malfunction of the instrument.
2. Carry the instrument with one hand grasping the arm and the other hand supporting the base.
3. Avoid bumping the microscope on the table or in the cabinet.
4. Always place the microscope squarely in front of you with the ocular facing you.
5. To prevent accidents, do not drape the cord over the edge of the table or into your lap.
6. Since the microscope is an electrical instrument, follow all safety rules concerning wiring and plugs.
7. Avoid touching or otherwise soiling the surfaces of the glass, lenses or objectives. When necessary, wipe the lens surfaces with clean **lens paper** only.
8. Keep the stage dry and clean. Use the stage clips or "claw" to hold the slide in place.
9. Never view a wet mount slide without a cover glass (slip).
10. Adjust the binocular head for one clear field of view or train yourself to keep both eyes open when looking through a monocular microscope.
11. When returning the microscope, make sure the lowest power objective is in position, the dust cover is placed back on and the cord is secured with a rubber band. **DO NOT WRAP THE CORDS AROUND THE BASE OF THE MICROSCOPE.**

Focusing the Microscope

1. First place the slide on the stage using the clips or stage mechanism and position the center of the cover slip directly over the condenser and stage opening.
2. Always start with a low light. A lightly stained slide or one too brightly lit will "bleach" the specimen and make it difficult to view. Alternatively, if the slide is too dark or darkly stained, adjusting the light upwards will brighten the subject.

3. Begin with the scanning or low power objective positioned all the way down with the course adjustment knob. Moving the stage, find any material and focus sharply with the fine adjustment knob.

4. Always center the specimen in the center of the field of view before moving to the next higher objective.

5. Note as the objectives progress upwards in magnification, the objectives also have a longer length. Care should be taken to adjust the scope accurately under the lower power before moving to the next higher objective. If correctly done, the object should be slightly visible under "high dry" or "high oil" power. **ONLY THE FINE ADJUSTMENT KNOB** should be used under this magnification.

6. If a sharp image is not possible, drop down to the lower magnification and repeat the steps again. Since microscopes are constructed to be **parfocal**, little or no adjustment should be necessary when progressing from low power to high power.

7. At each change of slides or magnification, light adjustment should be done with the light dial and/or the iris diaphragm control lever.

Laboratory Drawings

Drawing is one of the greatest aids to observation and the oldest method of recording observed specimens. Visual observation is an integral part of laboratory work. Before beginning to draw, make a thorough observation of the object. Laboratory drawings should represent something as it is, not imagined or embellished. Every line or dot should represent some fact or structure.

All drawings should be made with a 2-H drawing pencil. This pencil should have a sharp point at all times. Be sure to make the drawing large enough to include all details without crowding. Relative proportions are very important; e.g., the relation of the breadth to the length of the object. Draw only what can actually be seen, and do not attempt to draw more than you can represent properly.

Simple line drawings are usually better than shaded drawings and are much easier to make. The outline edges should be sharp, clear, and smooth. If shading is done, use very small, barely visible dots in a "stippling" effect.

Each drawing or photograph should have all of its essential parts labeled. Extend straight horizontal lines from the various parts with a ruler. Print the name of the parts at the end of the lines. All labels should be the same distance on the right-hand margin. If there is not enough room on the right without crowding, use the left-hand side.

Preparation of (Stained or Unstained) Wet Mounts

In producing a temporary wet mount, these steps should always be followed (Figure 1.1):

a. Place a clean slide on the table or desk.

b. Add a drop of water to the slide.

c. Place the specimen to be studied in the drop of water.

d. Cover the specimen or material with a cover glass (Be careful! These may break or scratch easily).

e. Eliminate air bubbles by placing the edge of the cover glass directly in the water and slowly decreasing the angle until the cover glass rests completely over the object.

f. If there is too much water for the cover glass, use a blotter to absorb the excess.

g. If specimens are already in a liquid, there is no need for additional drops of water.

h. To stain a wet mount, place a drop of dye or stain on one side of the cover glass. Place absorbent paper on the opposite side of the cover glass. The paper will draw the stain under the cover glass.

i. Clean and dry slides and cover glasses when finished. Return them to their proper place.

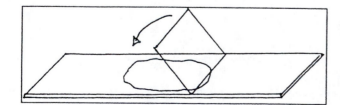 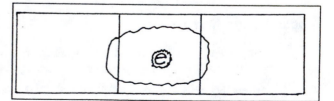

FIGURE 1.1 Wet Mount Preparation

Observation of Inversion

Using ordinary newspaper print and following the directions above, prepare a wet mount of the letter "e." Place the temporary slide on your stage and fasten with the clips or the "claw" of the mechanical stage. Obtain the correct amount of light by adjusting the iris diaphragm. The low-power objective should be underneath the body tube. Look through the ocular and run the tube up very slowly. The letter should be visible at some point. Use the fine adjustment to achieve the sharpest image. Move the slide around and note how this affects the movement in the eyepiece. If a higher magnification is desired, swivel the next objective into position. Since the instrument is nearly **parfocal**, little or no adjustment should be necessary. Remember to only use fine adjustment when the high power is in use.

1. How is the image seen in the eyepiece different from the actual slide preparation?

   _____

2. What happens to the image when the stage is moved to the left? When the stage is moved upwards?

   _____

3. Why wouldn't the letter "O" be used for this preparation?

   _____

Depth Exercise

Obtain a prepared slide of colored threads. There are three different colors of threads on each slide (Practice with different slides since the sequence of colors varies with each one). Adjust the light with the iris diaphragm. Notice that this exercise depends on the use of both coarse and fine adjustment knobs. Careful adjustment must be done in order to see the threads in depth (Plate 1.2).

1. Which colored thread is on the top? _____

2. Which colored thread is in the middle? _____

3. Which colored thread is on the bottom? _____

4. Does the microscope view distort or change the actual **position** of the threads (top to bottom)? _____

Magnification and Measurement

The total magnification of the image is determined by the lenses of the oculars and objectives. These vary in their ability to magnify objects. The total magnifying capability is equal to the product of the two systems. To find this value it is only necessary to multiply the individual magnification of the ocular and the objective in place; i.e., 10X multiplied by 10X equals total magnification. In this example, the magnification would be 100 times the size of the objective.

1. What would the total magnification be of an object using the scanning objective (4X)? _____

2. What would the total magnification be of an object using the "high dry" power (40X)? _____

PLATE 1.2   Silk Threads

3. What would the total magnification be of an object using the "high oil" power (100X)? _____

For accurate drawings with relative proportions, it is often necessary to measure the size of the specimen being observed. Micrometers are special optical instruments for this purpose and are often built in to the ocular portion of newer scopes. Position the object near this scale and calculate the size of the specimen. If the ocular does not have a measuring scale, an estimate of the viewing field and specimen size can be accomplished by the following procedure:

1. Place a transparent millimeter ruler over the condenser.
2. Lower the low-power objective into position to observe the lines or divisions.
3. Move the ruler so that one line is touching the edge of the left field of vision.
4. Count the number of lines or divisions between the left-and right-hand margins of your field of view.
5. Determine the approximate diameter of the field. What is the answer in millimeters? _____

In the Appendix is a reference table of the metric system. Note that 1000 millimeters equals one meter. Alternatively 1000 micrometers are equal to one millimeter. Most biological measurements are made in terms of micrometers.

It is also possible to calculate the field of view for the high- power objective by multiplying the *low-power field of view* by the *low-power lens magnification* and divide the result by the *high-power lens magnification*. For example, if the low-power field of view is 1.5 mm with the 10X objective lens, then the field of view for the 43X objective lens would be:

$$(1.5 \text{ mm}) (10X) \div 43X = 0.3488372 \text{ mm}.$$

6. Convert the diameter of the low power field of view from question # 5 above into micrometers.

_____

7. Calculate the diameter of the field of view for the high power objective using the values obtained in question # 5.

_____

Dissecting Microscope

With specimens or slides too large for the compound microscopes, a dissecting microscope allows magnifications from 5X to 50X. This type of scope generally has: a dual eyepiece which adjusts for different eye widths, two coordinated focusing knobs, one magnification dial near the ocular, and a mounting plate for specimens (Plate 1.3 A, B).

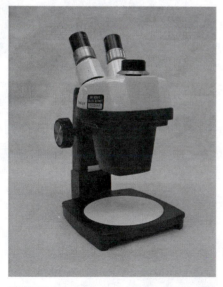

PLATE 1.3 A   Leica Stereo Zoom

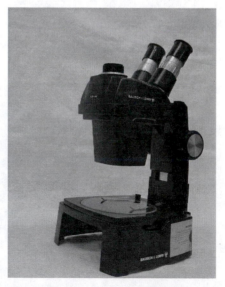

PLATE 1.3 B   Bausch & Lomb

Generally, if the mounting plate is clear there is a swivel mirror below which catches light from an independent lamp and reflects it to the specimen. If the plate is not clear, it will release and reverse from a light to dark surface. Select the best background for the greatest contrast in the sample.

For whole specimens, surface features can be observed and magnified by shining light directly down onto the top surface. Besides the magnified view, the object will also have a sharp, three-dimensional appearance.

Place a bird feather (or some other object) on the dissecting microscope. Move the object around and change the light and magnification. Next, place a prepared slide of a large animal cross section (Ex: earthworm cs). Again, move the mirror or change the colored plates for the best background and adjust the magnification.

Questions

1. How does a compound microscope differ from a simple microscope?

2. What does the "x" in 10x and 43x mean?

3. What is the function of the condenser?

4. What is the function of the iris diaphragm?

5. How is the total magnification of an object determined?

6. When making a wet mount, what procedure is followed to avoid trapping air bubbles under the cover glass?

7. Is the image seen in the dissecting scope inverted as in the compound microscope? (Hint: move the sample left/right)

8. Name some other biological specimens for which the dissecting microscope would be better than the compound microscope.

EXERCISE 2 Molecular Models

Learning Objectives

✔ Demonstrate the shape of some of the more important organic compounds with molecular models.

✔ Explain the difference between organic and inorganic compounds.

✔ Explain how covalent and ionic bonds differ from each other.

✔ Explain how carbon is capable of forming the variety of organic compounds found in living matter.

Introduction

Regardless of how simple or complicated animal or cell structures are, they consist entirely of chemical compounds. Four elements-*carbon, hydrogen, oxygen,* and *nitrogen*-make up approximately 95 percent of the weight of cell and animal composition. Organic compounds found in animals usually come from living things; plants and/or animals. **Organic** compounds may be defined as carbon containing compounds. All **inorganic** compounds found in animal cells are of mineral origin, either directly or indirectly.

An **element** is a substance composed of a single kind of atom. All matter in the earth is made up of approximately 103 elements. These are recognized by their short-hand symbols (Ex: C=carbon, Na=sodium, etc) and listed in a chemical periodic table. A **molecule** is a substance that is made up of two or more atoms while a chemical **compound** is a substance that contains two or more *different* elements.

An individual atom is composed of three fundamental parts called *protons, electrons and neutrons.* The positively charged particles or **protons** and the neutral particles, **neutrons** are located in the center of the atom or nucleus. The **electrons,** or negatively charged particles arrange themselves around the nucleus in shells or orbits. Charged atoms formed by the loss and gain of electrons are called **ions.** Ions are negatively or positively charged atoms. Particles with opposite charge are attracted to each other while those with the same charge are repelled by each other.

A **bond** exists whenever two or more atoms or ions become fixed. These particles do not touch each other, but are fixed at a very small distance by attractive forces. Two kinds of attractions are called **ionic bonding** and **covalent bonding.**

Since ions in a compound have opposite charges, they are attracted to one another. This type of bond is known as an ionic bond. Thus, an ionic bond is a force of attraction between oppositely charged particles. Ions can combine in different ratios. The combining capacity of an element is called its **valence.** Valences are represented by a number bearing a (+) or a (−) sign to represent the charges on the ion; e.g., hydrogen (H^+), oxygen (0^{-2}), and nitrogen (N^{-3}).

A second mechanism of fixing atoms can be accomplished by the sharing of electrons through a **covalent** bond. The electrons in the shared pair are furnished by separate atoms. A **single** bond is one pair shared between two atoms or cores and is represented by a single "dash" line between atomic symbols (Figure 2.1, A, B, C). Consequently, if two pairs of electrons are shared between atoms, it constitutes a **double** bond (Figure 2.1, D, E, F, G, H, I, J). Written as a double bar between atomic symbols, this bond is common between carbon and several other elements.

Procedure

For this laboratory exercise, molecular models will be provided. (Since various kits are available, follow the directions for the specific one provided in lab). Each group should have a set of "atoms"—colored balls. Only four colors of balls will be used (Black balls represent the carbon atom; yellow balls represent hydrogen, red balls represent oxygen, and the light-blue represent nitrogen). In the molecular boxes should also be sticks and flexible springs for "bonding".

The number of holes in each ball or "atom" represents the number of its available sites for bonding. Examine the hydrogen ball. It has only one hole for a single bond site while carbon has 4 holes or bonding sites. Using the key below and the drawings in Figure 2.1, connect various atoms to one another using the connecting sticks.

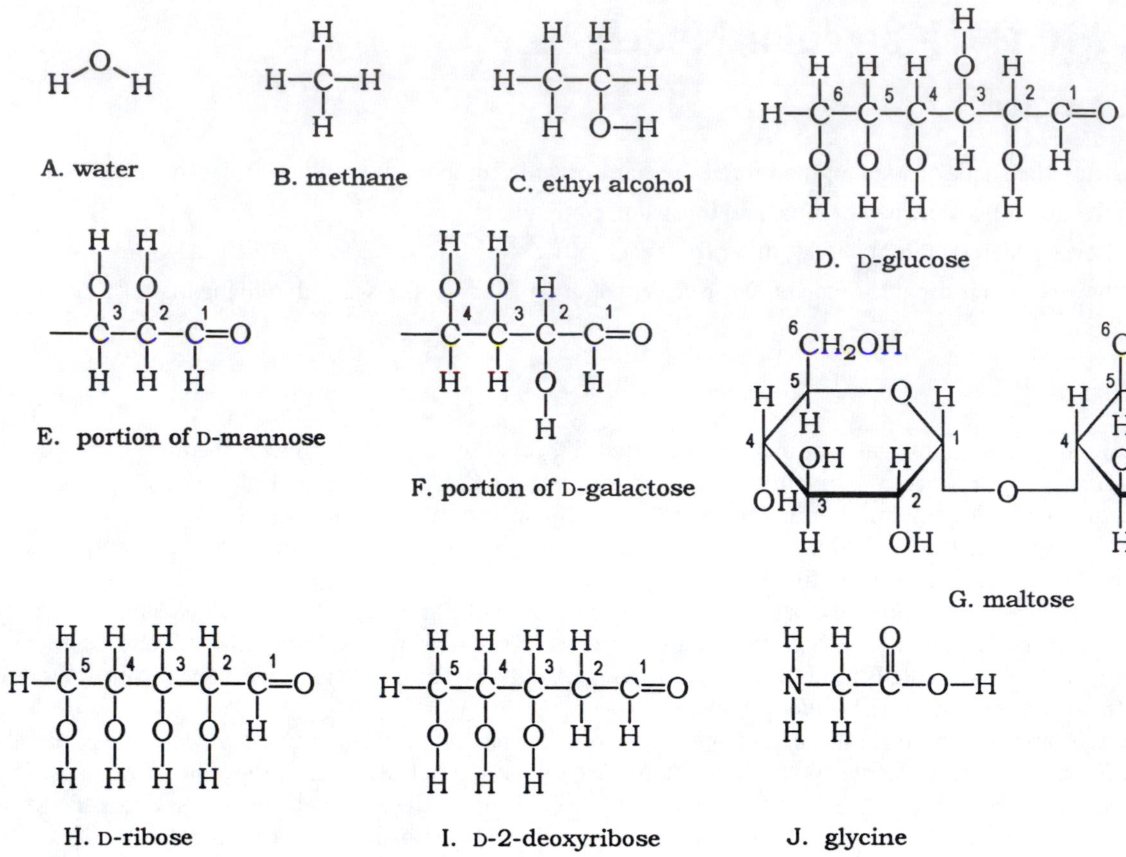

A. water

B. methane

C. ethyl alcohol

D. D-glucose

E. portion of D-mannose

F. portion of D-galactose

G. maltose

H. D-ribose

I. D-2-deoxyribose

J. glycine

FIGURE 2.1   Varied Biological Molecules

Bonds	Use	
long	carbon-to-carbon	(–C–C–)
short	hydrogen-to-carbon	(–C–H)
	hydrogen-to-oxygen	(–O–H)
	oxygen-to-carbon	(–C–O–)
	double bond	(–C=O)

Water Molecule

Make a water molecule (H_2O) with an oxygen ball, two hydrogen balls, and two short woods. Now separate one hydrogen atom from the molecule. The separated parts have produced a hydrogen ion (H^+) and a hydroxyl ion (OH^-). This occurs to a very limited extent in water (Figure 2.1 A).

Methane Molecule

Methane is the simplest hydrocarbon molecule (CH_4). Use a carbon ball, four hydrogen balls, and four short woods for this model (Figure 2.1 B).

Now make an ethane molecule (C_2H_6), a two-carbon hydrocarbon. Write out the structural formula for ethane in a similar manner to those shown in Figure 2.1. Remove one of the H atoms from ethane and substitute a hydroxyl ion (OH^-). This represents the compound known as ethyl alcohol (Figure 2.1 C).

D-Glucose Molecule

Several major organic compound groups are familiar as nutritional food groups. They are also the building blocks for more complex cellular compounds. The **carbohydrate** group is composed of building blocks called **simple sugars** or monosaccharides. One of the most important of these simple sugars is **d-glucose**. Make a molecule of d-glucose using the formula below (Figure 2.1 D).

Note the empirical formula, $C_6H_{12}O_6$. The empirical formula is partially important in determining chemical characteristics. The different possible arrangements of the atoms and groups around carbon atoms result in distinct differences in properties. There are several different $C_6H_{12}O_6$ molecules, or **hexoses**. Notice that the carbon atoms are numbered from 1 to 6. Reverse the positions of the H and OH on the number 2 carbon atom of the glucose molecule just assembled. This is **d-mannose**, a different simple sugar but possessing the same empirical formula as d-glucose (Figure 2.1 E).

Restore the model to d-glucose and then reverse the positions of the H and OH on the number 4 carbon. This is **d-galactose**, yet another monosaccharide containing 6 carbon, 12 hydrogen and 6 oxygen atoms (Figure 2.1 F).

Maltose Molecule

Another carbohydrate group is the **disaccharide** level. Composed of two simple sugars, these are larger in size and more complex. Restore the original glucose molecule. Remove the H atom and the OH on number 5 carbon and connect the remaining O atom to the number one carbon. Select an adjacent partner group and complete the following steps. Group two will remove the OH group on number four carbon and add an H atom to number one atom. Now join the two **pyranose** ring glucose molecules together, connecting the oxygen atom of group 1 to carbon 4 of group 2.

This is a model of the **maltose** molecule, a disaccharide consisting of two glucose residues (Figure 2.1 G). Note what is left over. There should be two hydrogens and one oxygen atoms remaining. This is an example of **dehydration** and can be continued to include dozens of glucose residues to form **polysaccharides** such as starch, glycogen, cellulose, etc. Separate the two glucose residues and insert the H and OH back into the original model. The splitting of a molecule by adding water is known as **hydrolysis**. This process is common in the digestive tract when large complex carbohydrates must be broken down (digested) into smaller simple sugars.

Pentose Molecule

Some simple sugars containing 5 carbons (**pentoses**) are useful as constituents of nucleic acids; the basic units of genetics. Two types of **nucleic acids** which contain pentoses are deoxyribonucleic acid, or **DNA**, and ribonucleic acid, or **RNA**. Make the 5-carbon pentose sugar **d-ribose** as given in Figure 2.1 H. Now, replace the OH from the number 2 carbon and replace it with an H atom forming **d-2-deoxyribose** (Figure 2.1 I). *Deoxy* means "minus one oxygen," and one oxygen is the only difference between the two five-carbon sugars. These sugars represent just a portion of the larger DNA and RNA molecules.

Glycine Molecule

Glycine is the simplest **amino acid**. Amino acids are the substances from which the organic compounds, **proteins**, are synthesized. Notice that this molecule has another element in addition to carbon, hydrogen, and oxygen. Make a glycine model according to the formula (Figure 2.1 J).

Questions

1. How many bonds can carbon form with other elements?

2. What elements make up most of the weight of an organism?

3. Name the three subatomic particles that make up an atom.

4. How are these subatomic particles arranged in the atom?

5. How do inorganic and organic compounds differ?

6. What are the important classes of organic compounds?

7. What organic units are commonly used to form proteins?

8. What chemical reaction is used to join organic compounds?

9. Give some of the characteristics of water.

EXERCISE 3   Cytology and Histology

Learning Objectives

✔ Identify the organelles of the cell and their functions.
✔ Explain the difference between tissue, organ, and organ system level of organization.
✔ Recognize and identify the various types of tissues and their functions.
✔ Distinguish between prokaryotic and eukaryotic cells.

Introduction to Cell Structure

One of the basic principles of biology states that all living things are composed of structural and functional units called cells (Figure 3.1). Animals are composed of these same units and their study is called cytology. Whether the organism is composed of one cell or many, a cell is the structural unit of its composition. Functionally, the cell has the ability to assimilate, divide, mature, differentiate, reproduce, and respond to stimuli.

In the development of multicellular organisms, individual cells begin to cooperate and specialize. This essentially results in groups of cells with specific functions called tissues and the study of histology. Many different kinds of cells and tissues will be observed in the laboratory. The prokaryotes, represented by the bacteria forms, are cells which lack a "formed" nucleus and membrane-bound organelles. The other category of cells, the eukaryotes, compose the typical animal body and will be studied here.

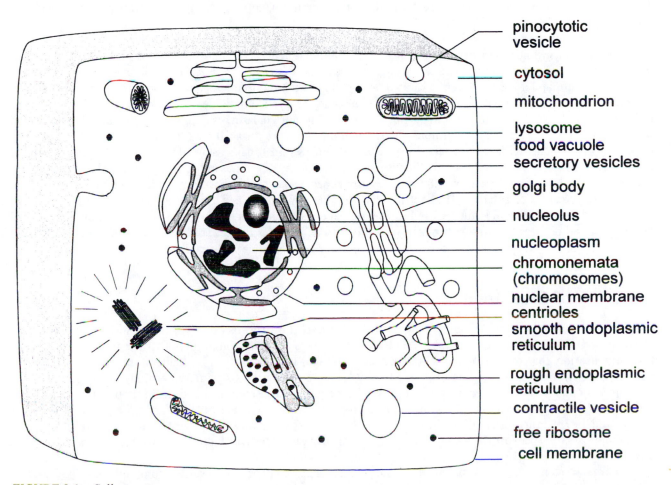

FIGURE 3.1   Cell structure.

All eukaryotic animal cells are composed of similar parts. The **cell membrane** is the living outer boundary of the cytoplasm. It is semipermeable and therefore controls the flow of materials in and out of the cell's interior. The membrane is very active, especially in the processes of **phagocytosis, pinocytosis,** and **membrane transport.**

The **cytoplasm** proper is a fluid, mobile, yet granular matrix. Its living components are called **organelles** (little organs) while **inclusions** are the nonliving components. *Mitochondria, ribosomes, vacuoles, Golgi bodies, lysosomes, centrioles* and the *endoplasm reticulum* are the major organelles found in the cytoplasm proper. A brief description and function of each organelle is given.

Mitochondria appear as granules or long rods and possess two membranes. While the outer membrane is smooth, the inner membrane folds inward into transverse shelves called *cristae*. These structures perform the process of cellular respiration by furnishing an energy-rich compound, *adenosine triphosphate* (*ATP*), to the cell. Mitochondria also have *DNA* and *RNA* within their matrix.

The **endoplasmic reticulum (ER)** is a convoluted network of membranous, weaving channels through the cytoplasm. Its external surface may be agranular or granular. Granular (rough) endoplasmic reticulum is studded with ribosomes while the agranular (smooth) lacks the ribosomes. The endoplasmic reticulum is at places continuous with the plasma membrane and the nuclear membrane. The ER synthesizes various lipids, proteins and carbohydrates required by the cells or destined to enter the Golgi complex for further processing.

Ribosomes lie free in the cytoplasm or are attached to rough endoplasmic reticulum. They are very fine granules composed of protein-bound ribonucleic acid (RNA). Protein synthesis is the major role of ribosomes. They are unusual cell components because they lack a membrane.

Golgi bodies appear as tiny parallel vesicles that form from outpockets of the endoplasmic reticulum. Research data has shown that proteins formed by ribosomes on the surface of the rough endoplasmic reticulum are channeled to the cisternae of smooth endoplasmic reticulum. There they are packaged in *transition vesicles*. The transition vesicles then move to the Golgi bodies where the proteins are chemically modified, sorted, and packaged into vacuoles. Some vacuoles move to the cytoplasm and become known as **lysosomes.** Other vacuoles join with the cell membrane and release their contents to the intercellular space. Lysosomes are spherical in shape with little or no internal organization. They contain enzymes for digestion. Oil droplets, granules, nutritive substances, and vacuoles make up the various types of **cell inclusions.**

The **nucleus** is the most prominent part of the cell. There is usually one nucleus per cell although specialized cells may lack a nucleus while others may be binucleated or multinucleated. It is composed of four distinctive parts: *nuclear membrane, nucleoplasm, chromosomes* with DNA, and one or more *nucleoli*. The **nuclear membrane** is the outer boundary of the nucleus and it has two layers separated by a space. *Pores,* tiny circular openings, occur along the membrane. The outer nuclear membrane is also continuous with the endoplasm reticulum. The membrane controls the flow of materials between the ground cytoplasm and the nucleus proper. The **nucleoplasm** is a clear matrix, similar to the cytoplasm which contains the chromosomes and nucleoli. The **chromosomes** are linear, thread-like structures composed of the hereditary material, deoxyribonucleic acid (DNA). The **nucleolus** is a small, darkly stained area within the nucleus proper. Because it is the site of RNA synthesis and the ultimate source for the ribosomes in the cell, there may be one or several active nucleoli within any particular cell. The **centrosome** occupies a central position within a cell near the nucleus. Within this mass are two centrioles. The centrioles play a very prominent role in the process of mitosis or cell division.

Although the diagram in Figure 3.1 indicates a typical eukaryotic, animal cell with its component parts, do NOT expect to see all of these features within cells using the compound microscope. This diagram represents an idealized cell as pictured with an electron microscope at magnifications not possible with the typical student microscope.

Hierarchy of Organization

Within the realm of the animal kingdom, different complexities of organisms exist. The lowest forms exhibit a single cell or **unicellular** form. In higher animals, such as man, there is a greater complexity of structure that requires further organization and coordination of metabolic activities. This is accomplished by cooperation of cells into coordinated **tissues.** In many animals, the origin of these tissues can be traced to primary **germ layers** organized in the developing embryo. The outer **ectoderm** forms certain integumentary structures and nerve tissue while the middle **mesoderm** forms blood, bone, cartilage, and muscle tissues. The inner **endoderm** layer forms the epithelial lining of the gut and the structures arising from it; e.g., liver and lungs. Just as cells are grouped into tissues, tissues are grouped into recognizable shaped forms with specific functions called **organs.** The liver, kidneys, stomach, pancreas and gonads are some well-known organs.

An **organ-system** level is accomplished by coordination of groups of organs towards a common goal. The *digestive system, circulatory system, excretory system, nervous system,* and *reproductive system* are common examples of these sophisticated levels found in higher organisms. The remainder of this exercise will be devoted to the study of cells and tissues while organs and organ-systems will be studied in later exercises.

Histology

Common tissues forming the sophisticated animal body are generally divided into four general categories. These divisions are: epithelium; connective; muscle and nervous while reproductive cells such as sperm and egg are considered a "specialized" category.

Epithelial Tissue

Epithelial cells form a compact, continuous layer covering the surface of the body or lining cavities leading to the exterior of the body. These cells form a protective layer and/or have a secretory or absorptive role. Several subcategories are recognized dependent upon: (1) shape of cell; (2) number of cells in the epithelial layer and (3) presence of cilia. Three general shapes of epithelial are defined as: **squamous, cuboidal** and **columnar.** A single layer of epithelial cells, free at one surface and attached at the other is called a **simple** layer while layered tiers of epithelial cells are called **stratified** (Figures 3.2, 3.3).

Epithelial Tissue: Subcategories

1. *Simple Squamous.* A single layer of thin, flattened cells that line the heart and blood vessels, alveoli of the lungs, labyrinth of the inner ear, lining of serous membranes, Bowman's capsule, loop of Henle, and the lens of the eye (Figure 3.2 A, Plate 3.1 A).
2. *Simple Cuboidal.* A single layer of cube-shaped cells that line the ducts of many glands, such as the thyroid, surface of the ovary, and tubules of the kidney (Figure 3.2 B, Plate 3.1 B).
3. *Simple Columnar.* A cell type that is taller than it is wide and whose nuclei are usually next to the attached edge of the cell (Figure 3.2 C, Plate 3.1 C). Simple columnar lines the digestive tract from the lower esophagus to the anus, gall bladder, and lines the excretory ducts of many glands. Some cells, known as **goblet cells,** produce mucus and are interspersed among these cells.
4. *Ciliated Columnar.* The general structure of this cell is like that of simple columnar except that the free edge of the cell has cilia. This tissue lines the nose, fallopian tubes and uterus, and the small bronchi.
5. *Pseudostratified Ciliated Columnar.* This cell type is a columnar cell with cilia, but whose nuclei are arranged in layers making the tissue appear to be stratified although it is a single layer of cells (Figure 3.2 D, Plate 3.1 D).
6. *Stratified Squamous.* A multiple layer of cells forms this epithelium that varies in shape from rounded cells at the base of the tissue layer to squamous cells at the free edge (Figure 3.3 A, Plate 3.1 E). This tissue forms the outer layer of the skin where it has a cornified (keratinized or horny) layer on the free edge. A noncornified variety lines the mouth and esophagus. It also forms a portion of the following structures: cornea, conjunctiva, epiglottis, and vagina.
7. *Transitional.* This tissue is found only in the walls of the urinary bladder and ureters. Like the stratified squamous, it is a layered form although the outermost cells are rounded rather than flattened. This allows for the distention and stretching that occurs in the filled bladder and ureter tubes (Figure 3.3 B, Plate 3.1 F).

Connective Tissue

Connective tissues bind and support other tissues. Each tissue type consists of three elements: a cell, fiber, and matrix. The cell secretes the fiber and surrounding matrix. Three types of fibers are found in connective tissue: white (collagenous), yellow (elastic), and reticular (branching). Most connective tissues consist of relatively few cells and a great deal of fiber and/or matrix. Variations in all three components can produce a wide variety of connective tissue types (Figure 3.4).

Connective Tissue: Subcategories

1. *Areolar.* This tissue is a loose, irregular grouping of white and yellow fibers with a large amount of matrix and cells (Figure 3.4 A, Plate 3.2 A). Areolar tissue is present in many body sites and includes mesenteries, basement membrane of skin, and organ capsules.

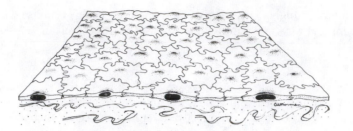

Figure A. Simple Squamous

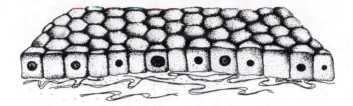

Figure B. Simple Cuboidal

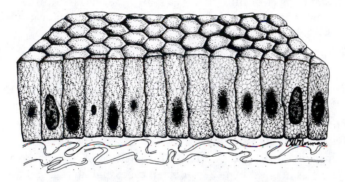

Figure C. Simple Columnar

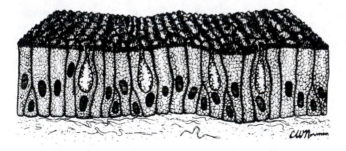

Figure D. Pseudostratified Columnar

© Kendall/Hunt Publishing Company

FIGURE 3.2 Nonstriated Epithelial Tissue

2. *White Fibrous* (*Collagenous*). White fibrous is primarily a compact, parallel arrangement of collagenous fibers with little matrix and cells called **fibroblasts.** This tissue forms tough tendons, ligaments, and sheaths (Figure 3.4 B, Plate 3.2 B).

3. *Yellow Fibrous* (*Elastic*). Like white fibrous tissue, yellow tissue is also a compact, parallel arrangement of fibers, but composed of elastic fibers. Yellow fibrous ligaments are found along the vertebral column.

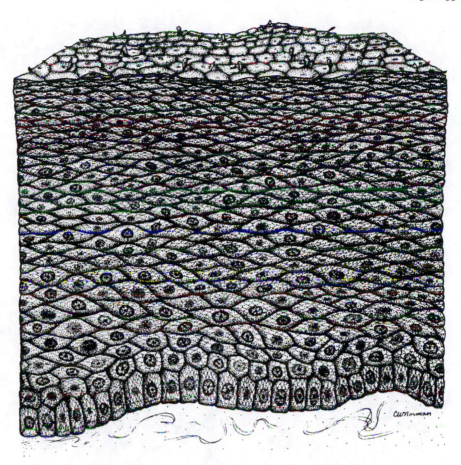

Figure A. Stratified Squamous Epithelium

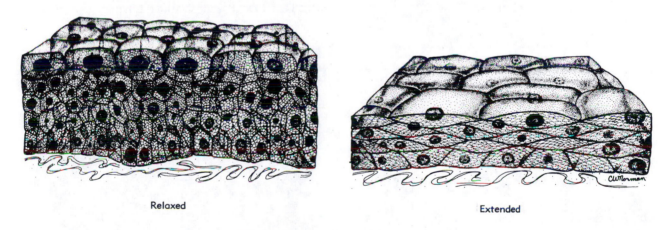

Relaxed

Extended

Figure B. Transitional Epithelium

FIGURE 3.3   Stratified Epithelial Tissue

4. *Adipose Tissue (Fat)*. Adipose tissue consist of large rounded cells (**adipocytes**) filled with oil whose nuclei and cytoplasm are squeezed to one side of the cell (Figure 3.4 C, Plate 3.2 C). Fat tissue may be found beneath the skin, at the edge of the kidney, between skeletal muscle, and in depots such as the buttocks and abdomen.

5. *Hyaline Cartilage*. Cartilage is primarily a matrix of pliable, densely arranged parallel protein fibers (Figure 3.4 D, Plate 3.2 D). Hyaline cartilage is a clear variety of cartilage found in the nose, larynx, trachea, and ends of bones. **Fibrocartilage** has collagenous fibers embedded within it and forms the intervertebral disc of the vertebral column. **Elastic** cartilage has elastic fibers embedded within it and forms the support for the ears.

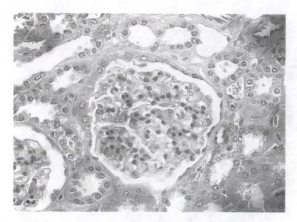

PLATE 3.1 A   Simple Squamous

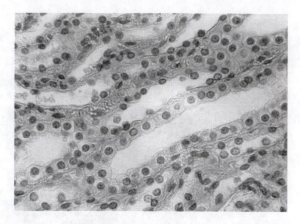

PLATE 3.1 B   Simple Cuboidal

PLATE 3.1 C   Simple Columnar

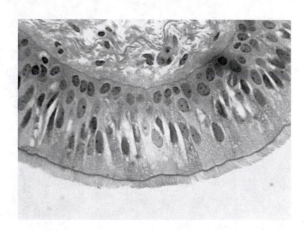

PLATE 3.1 D   Pseudostratified Ciliated

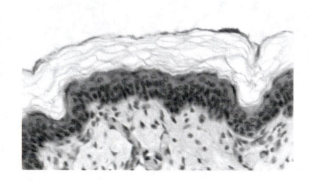

PLATE 3.1 E   Stratified Squamous with Keratin

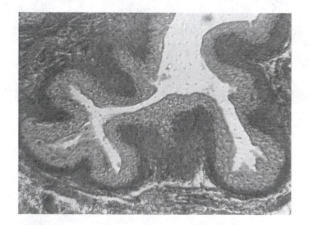

PLATE 3.1 F   Transitional

6. *Bone.* Like cartilage, bone is mostly matrix. The matrix, however, is a hardened complex of calcium phosphate and calcium carbonate. The microscopic structure of dense bone shows that it is composed of units known as **Haversian systems** or **osteons.** Each Haversian system contains channels called *Haversian canals* which contain blood vessels and nerves surrounded by concentric tubes of bone (Figure 3.5, Plate 3.2 E). Bone cells (mature **osteocytes**) that form these tubes of bone become trapped in spaces known as **lacunae.** Small channels of **canaliculi** allow communication between different layers of bone cells.

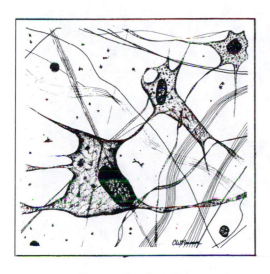

Figure A. Areolar

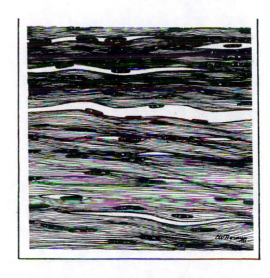

Figure B. Dense Fibrous

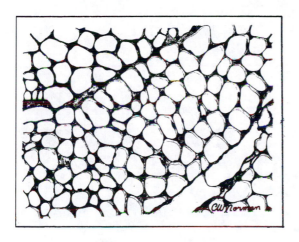

Figure C. Adipose

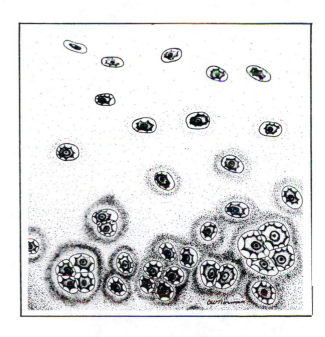

Figure D. Cartilage

FIGURE 3.4 Connective Tissue Types

7. *Blood.* Blood is a unique connective tissue because the matrix between the cells is primarily water and protein known as blood **plasma** (Figure 3.6, Plate 3.2 F). The majority of cells are *red blood cells* (erythrocytes) which carry oxygen and have no nuclei. A second type is the *white blood cell* (leucocyte) which is a defensive cell that has a distinctive nucleus. There are about 700 erythrocytes for each leucocyte. Lastly, a third type of fragmented cell, the *platelet,* constitutes a small percentage of the total formed elements. Platelets or **thrombocytes** are involved in blood clotting.

PLATE 3.2 A   Areolar

PLATE 3.2 B   White Fibrous (Tendon)

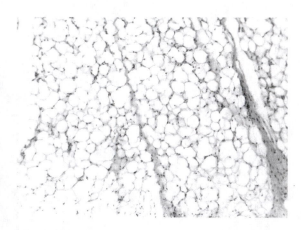

PLATE 3.2 C   Adipose

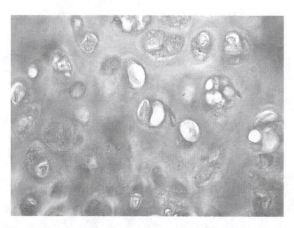

PLATE 3.2 D   Cartilage (Hyaline)

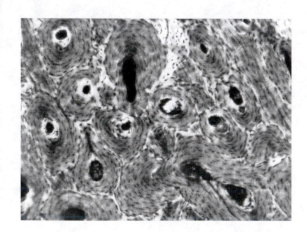

PLATE 3.2 E   Compact Bone

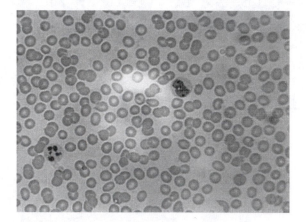

PLATE 3.2 F   Blood Tissue

Muscle Tissue

Muscle cells are elongated cells (fibers) that form the bulk of the body and are specialized for contraction. These cells may line ducts and propel fluids through the body or regulate the position and movement of various body parts. They are divided into categories dependent upon their physical appearance and nervous system control (Figure 3.7).

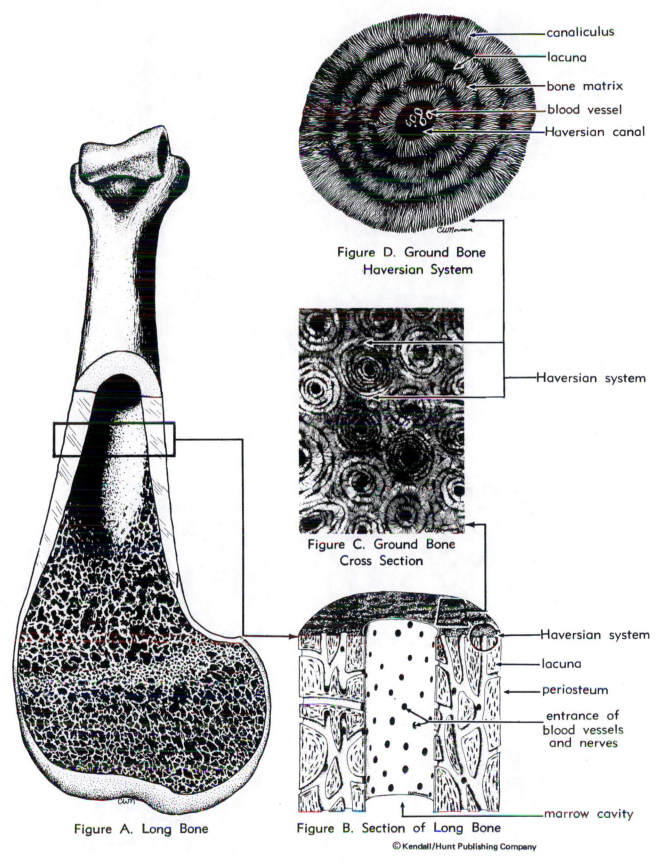

canaliculus

lacuna

bone matrix

blood vessel

Haversian canal

Figure D. Ground Bone
Haversian System

Haversian system

Figure C. Ground Bone
Cross Section

Haversian system

lacuna

periosteum

entrance of
blood vessels
and nerves

marrow cavity

Figure A. Long Bone

Figure B. Section of Long Bone

FIGURE 3.5   Bone Tissue

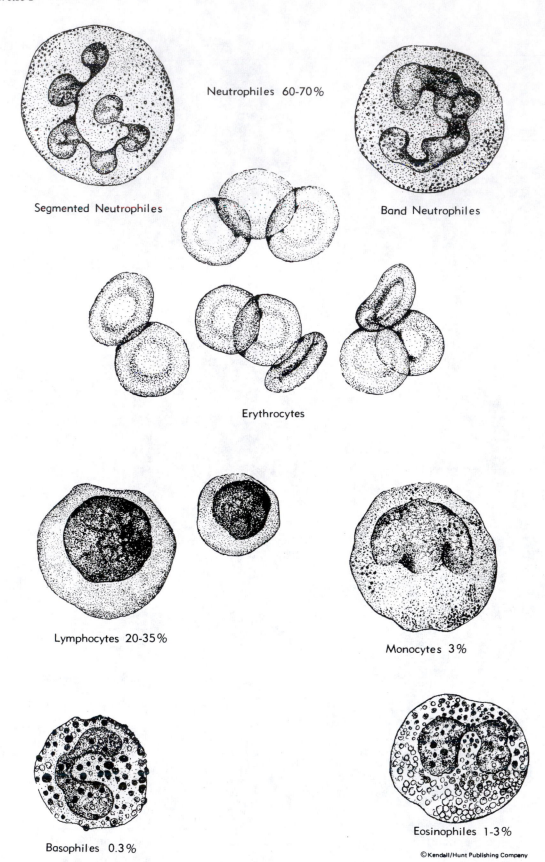

Neutrophiles 60-70%

Segmented Neutrophiles

Band Neutrophiles

Erythrocytes

Lymphocytes 20-35%

Monocytes 3%

Basophiles 0.3%

Eosinophiles 1-3%

FIGURE 3.6   Types of Blood Cells

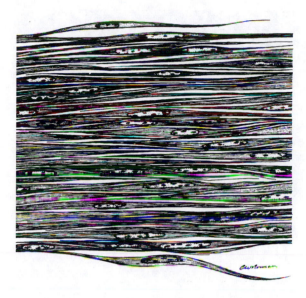

Figure A. Smooth Muscle

Figure B. Striated Muscle

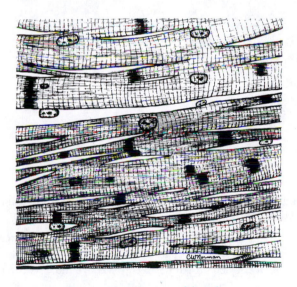

Figure C. Cardiac Muscle

FIGURE 3.7 Muscle Tissue Types

Muscular: Subcategories

1. *Smooth Muscle* (*Involuntary*). Smooth muscle fibers are elongated, spindle-shaped cells which have a single nucleus located centrally in the thickest portion of the cell (Figure 3.7 A, Plate 3.3 A). This cell type is found in the walls of blood vessels, dermis of the skin, uterus, digestive tract, urinary and reproductive tracts. It is recognized by the LACK of banded stripes within the cell. Because these contract without conscious effort, they are identified as involuntary muscle.

2. *Striated Muscle* (*Voluntary, Skeletal*). These cells are long, cylindrical cells that are multinucleated and have regular **striations** or bands. The cell maintains a uniform diameter throughout its length (Figure 3.7 B, Plate 3.3 B). The term striated refers to the transverse markings of the myofibrils. The arrangement of the my-

PLATE 3.3 A   Smooth Muscle

PLATE 3.3 B   Striated Muscle

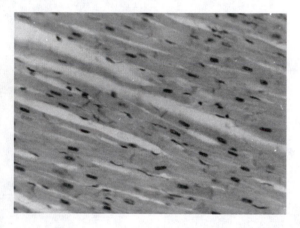

PLATE 3.3 C   Cardiac Muscle

ofibrils produces light and dark bands. These bands are noticeable in longitudinal sections. This tissue is attached to the skeleton and provides voluntary control over positioning of the body.

3. *Cardiac Muscle (Involuntary)*. Cardiac muscle consists of striated, cylindrical cells which are uninucleate and branched (Figure 3.7 C, Plate 3.3 C). These cells are much shorter than striated muscle cells. **Intercalated discs**, which are boundaries of adjacent cells form dark, transverse bands. As the name implies, this tissue is found only in the heart. The banding, intercalated discs, and branching are best seen in longitudinal sections of the slide. Since the heart has its own regulatory contractive control, this muscle type is also categorized as involuntary.

Nervous Tissue

Nervous tissue is composed of **neurons** (nerve cells) that are arranged in chain formation. Each neuron consists of a **cell body** with the nucleus; one or more branching extensions or **dendrites** that transmit impulses toward the cell body; and an **axon** that transmits impulses away from the cell body (Figure 3.8, Plate 3.4). These cells are found in the spinal cord, brain, spinal nerves, and cranial nerves. Other nerve cells which wrap, anchor and fill in spaces of the nervous system are **glial** cells. One of these specialized glial cells is the **Schwann** cell that wraps and protects the axons of nerves outside the brain and spinal column.

Reproductive Tissue

This specialized group of tissue consists of sperm produced in the seminiferous tubules of the testes and eggs produced by the ovaries. In the higher animal groups, sperm are small, mobile cells with flagella while eggs are usually large,

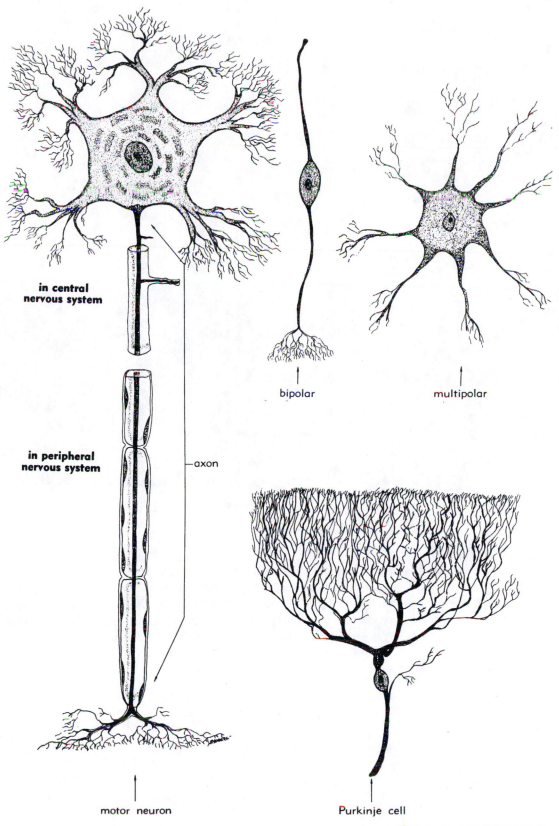

in central
nervous system

in peripheral
nervous system

axon

bipolar

multipolar

motor neuron

Purkinje cell

FIGURE 3.8 Nerve Tissue

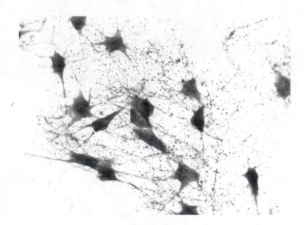

PLATE 3.4   Multipolar Neurons

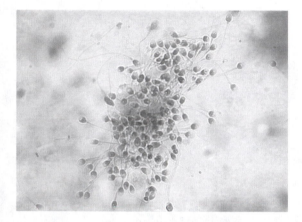

PLATE 3.5 A   Sperm

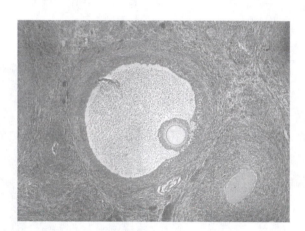

PLATE 3.5 B   Graafian Follicle

nonmotile cells which are mostly yolk (fat) (Figure 3.9, Plate 3.5 A, 3.5 B). Sperm from various mammals, such as rats or humans, have similar yet different characteristics. The cell is generally composed of a **head, neck** and **tail** segments. However, variations in the head region, enclosing the nuclear material, help to distinguish different sperm sources. Mammalian ovaries possess specialized structures called **Graafian follicles** to enclose and develop the female eggs. As each follicle reaches maturity, it is composed of protective cells (**corona radiata**) which surround the **oocyte** (egg) as it goes through meiosis. These in turn are embedded in a fluid filled cavity or **antrum** surrounded by protective **follicu-lar** cells.

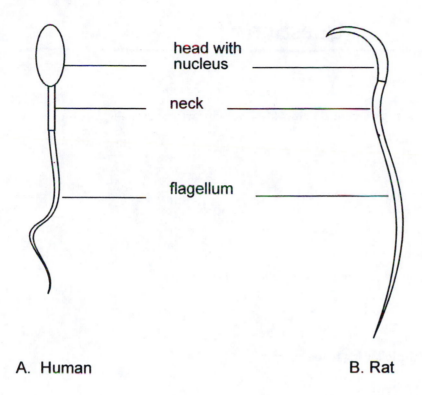

head with
nucleus

neck

flagellum

A. Human

B. Rat

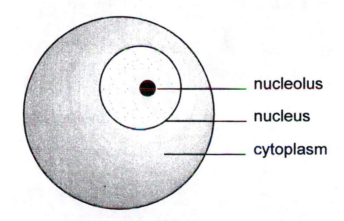

nucleolus

nucleus

cytoplasm

C. Starfish Egg

FIGURE 3.9   Reproductive cells: sperm (A-B) and egg (C).

Questions

1. The cell is the smallest structural and functional unit of life. What characteristics of life does the cell express?

2. Distinguish between prokaryotic and eukaryotic cells.

3. What is the structure and function of the various eukaryotic cell membranes?

4. What is the function of each of the following tissue types:

   epithelial?

   muscular?

   reproductive?

   connective?

   nervous?

5. Where in the body might you find examples of each of the tissue types in question #4?

Unit II
Division, Reproduction and Development

Cell Division

Learning Objectives

✔ Explain the significance of mitosis in reproduction and growth.
✔ Describe the different time phases of an interphase cell.
✔ Describe and recognize the four basic stages of cell division.
✔ Distinguish between karyokinesis and cytokinesis activity.

Introduction

Since no cell is immortal, a process is required by which they can be replaced. Whether cells function as either single individuals (such as the protozoans) or as a member of a group of cells (like yourself), the genetic instructions for directing the metabolic activities of the cell must be duplicated and passed onto new cells. The genetic instructions are encoded in a chemical called **deoxyribonucleic acid** (DNA). A **gene** is a section of DNA that specifies a protein or a trait. The number of genes that are required to make a functional organism varies according to the species. The number of genes required to make a fruit fly is estimated to be about 10,000 while that required to specify a human is approximately 100,000. To facilitate the management of these large numbers of genes during cell division, they are linked together in linear forms known as **chromosomes**. This mass of chromosomes, known as **chromotin**, exists as a diffuse network of extended or uncoiled chromosomes within the nucleus. This arrangement of genetic material gives the nucleus a darkened appearance in stained preparations of cells. The process by which somatic (nonsex) cells duplicate themselves is called **mitosis**. This mechanism precisely duplicates and apportions the chromosomes of one cell into two different daughter cells. Although mitosis is a continuous process, four basic stages are recognized: **prophase, metaphase, anaphase,** and **telophase**. The result is a continual and exact duplication of the genetic instructions—billions of times. The steps involving the actual division of genetic material are known as **karyokinesis** while the physical separation of the cell cytoplasm into two distinct cells is called **cytokinesis**.

Prior to mitosis, the chromosomes must be duplicated before they can be apportioned to daughter cells. This event occurs in **interphase**. Since the genes cannot be accessed for instructions during duplication, interphase must consist of a time period when genes can be accessed for their information as well as be duplicated. Thus, interphase, like mitosis, consists of several distinct phases itself. The first phase, G_1 (G for gap), is a period of time in which the cell is metabolically most active. The molecules required for growth, maintenance, and repair are produced. The second stage, the S phase (S for synthesis), is that time during which DNA is synthesized causing the chromosomal material to be doubled in number. The third phase, G_2, occurs just before mitosis. Ribonucleic acid (RNA) and proteins required for cell division are produced during this time.

Mitosis

Obtain a prepared slide of whitefish blastula. This slide has several groups of cells (embryos), each of which has many cells undergoing mitosis (Figure 4.1 and Plate 4.1). Find an individual cell in each of the stages by carefully examining the nuclear content.

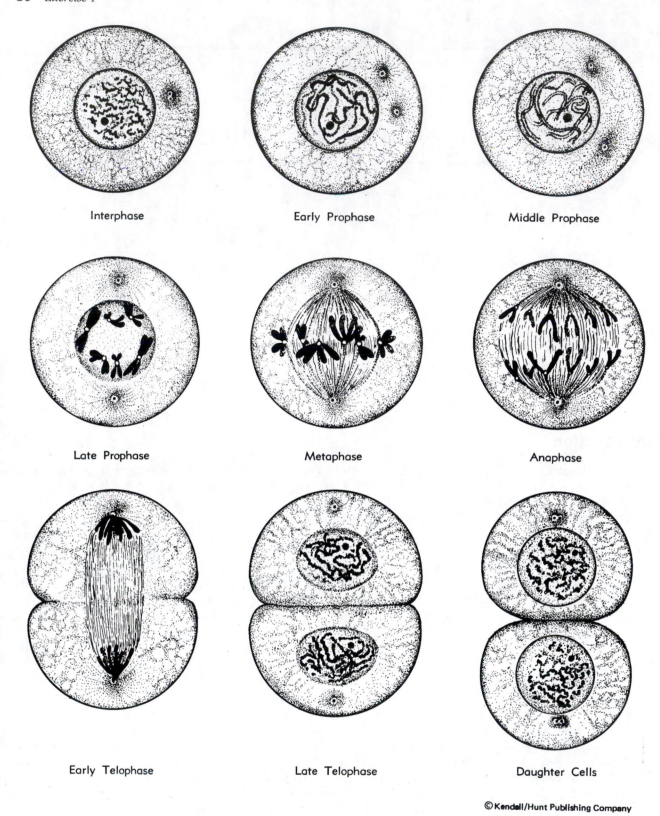

Interphase

Early Prophase

Middle Prophase

Late Prophase

Metaphase

Anaphase

Early Telophase

Late Telophase

Daughter Cells

FIGURE 4.1   Mitosis in animal cells.

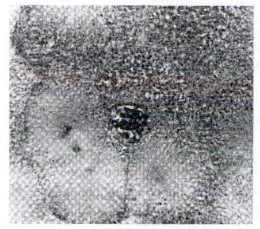

A. Prophase

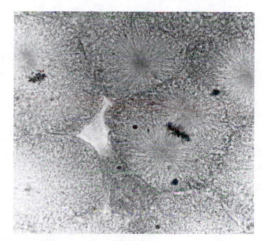

B. Metaphase

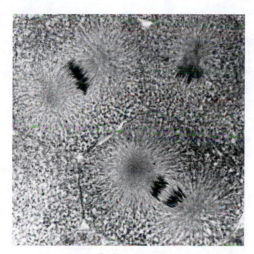

C. Anaphase

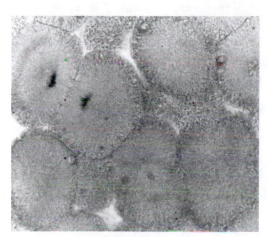

D. Telophase

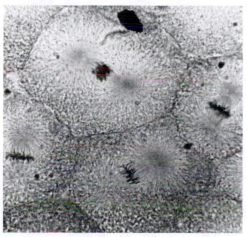

E. Identify Each

PLATE 4.1   Animal Mitosis.

Interphase

This "nondividing" stage or possibly preparatory stage is recognized by "normal" looking cells with a nondistinct nucleus. Cells in interphase will have a rounded nucleus surrounded by a defined nuclear membrane. Because the chromosomes are not compacted or condensed, the nucleus will appear lightly "stippled" and thread-like chromosomes will NOT be visible.

Prophase

This stage marks the beginning of mitosis. The centrioles divide and begin to move to opposite poles. The individual chromosomes begin to appear thicker and shorter due to coiling and uncoiling of chromatin. Because of the doubling of chromosomes which occurred in the S phase of interphase, each chromosome actually consists of two identical copies of the original chromosome, called **chromatids**, that are attached at their **centromeres**. The nuclear membrane will also begin to disintegrate during this phase as it allows for freer movement of the chromatids. The nucleolus also becomes inactive and disintegrates during this time.

Metaphase

This stage is marked by the arrangement of the chromosomes along the equator or center of the cell which is why mitosis is sometimes known as an "equatorial" division. The nuclear membrane and nucleolus have become disrupted. The centrioles have completed the migration to the poles and astral rays appear around each to form a unit known as an **aster** body. **Spindle fibers** extend out from the centrioles and attach to the centromeres of each chromosome.

Anaphase

The protein holding the centromeres of each chromatid dissolves and the chromatids can be considered separate chromosomes. The chromosomes now begin a rapid migration toward each centriole. As the chromosomes are migrating, they assume a "V" shape due to the trailing of each arm. The spindle fibers that are attached to centromeres contract or shorten at the centriole causing this movement. Late anaphase is marked by the grouping of the chromosomes into clusters.

Telophase

This final stage is marked by the infolding or furrowing of cytoplasm between the two daughter nuclei and a gradual pinching apart each of the two new cells. During this process, the new nuclear membranes appear and enclose each daughter nucleus. Nucleoli reform while the chromosomes uncoil and again appear as the chromatin network of normal cells.

Karyokinesis and Cytokinesis

The actual process of cell division is also separated into two time frames. **Karyokinesis** involves those stages (I, P, M, A, T) during which activity in the nucleus leads to the production of two distinct nuclear packets. **Cytokinesis** begins in telophase with the movement of the cell membrane inwards, marking the **cleavage** process, resulting in two physically separated daughter cells.

Questions

1. What are the hereditary structures found on chromosomes that control the expression of traits?

2. Of what chemical are these hereditary structures composed?

3. Where in the human body does mitosis ordinarily occur?

4. Describe the different stages of mitosis.

5. Why are spindle fibers important in cell division?

6. What are the different phases of interphase and what events occur during these times?

7. Why does duplication of the genetic material have to occur in interphase?

8. Identify the activities which distinguish karyokinesis and cytokinesis.

EXERCISE 5 Meiosis and Gametogenesis

Learning Objectives

✔ Describe the different stages of sex cell formation.

✔ Explain how and why the chromosome number is reduced to one half the normal complement during meiosis.

✔ Describe how traits are segregated during meiosis and explain how this segregation of traits contributes to genetic variation in the population.

Introduction

Since no cell is immortal and the environment can change with time and locality, a method of reproduction is needed that will allow the next generation to function as well as the parent generation did in a stable environment or produce the genetic variation needed to cope with a changing environment.

This variation is best achieved by combining the hereditary material from the nuclei of two individuals in fertilization. However, this fusion of cells would increase the amount of genetic material (DNA) per cell with each generation. Eventually, there would be no room left in the cell for chemicals other than DNA. Therefore, cells need a mechanism whereby the amount of DNA can be maintained at a constant level from generation to generation. This preparation for sexual reproduction is achieved with **meiosis** or **reduction division**.

Each cell has **genes** that direct the cell's metabolism. These genes are so numerous that they are linked together to form lengthy, string-like **chromosomes**. It is easier to manage a few chromosomes than thousands of individual genes during reproduction. For each chromosome in a eukaryotic cell, there is another chromosome that has the same linear sequence of genes. Therefore, chromosomes exist as *pairs* within cells and are said to be **homologous**. These matched chromosomes resulted from the previous fertilization which produced the organism, receiving one set from the father and one set from the mother. Chromosomes that do not have the same sequence of genes are referred to as **nonhomologous**.

Since gametes (sex cells) possess half the normal chromosome complement, one chromosome from each pair of homologous chromosomes, they are said to be **haploid** (one set). The cells from which gametes are produced are **diploid** cells (two sets of chromosomes). Thus, meiosis involves the change from a diploid state to that of a haploid condition.

Basically, meiosis is a process in which homologous chromosomes are duplicated and separated twice into four different cells when the cytoplasm divides. This reduces the chromosome number to one-half and segregates traits in the newly formed daughter cells. These daughter cells are called **sex cells** or **gametes** and are produced in specialized reproductive organs or **gonads**. In the male, the organs are called **testes**, while the **ovaries** produce the female gametes. Maturation of the gametes, **gametogenesis**, occurs in these same sex organs. The process of **spermatogenesis** involves the maturation of *spermatozoa* and **oogenesis** involves the maturation of *ova* (eggs).

Spermatogenesis is the entire maturation process of spermatozoa (haploid cells, sex cells and/or gametes) from a **primordial germ cell** (diploid) located in the testes. Oogenesis is the entire maturation process of ova (haploid cells, sex cells, and /or gametes) from a primordial germ cell (diploid) in the ovaries. Gametogenesis, or meiosis, involves: (1) production of four haploid nuclei from one diploid nucleus, (2) random segregation of traits by separating homologous chromosomes, and (3) recombination of genetic material. Recombination is due to **crossing over** between adjacent segments of homologous chromosomes during the *tetrad* stage of the first meiotic division (Prophase I). When these sex cells fuse in **syngamy**, considerable variation can be achieved in the offspring.

Meiosis

Obtain a prepared slide of *Ambystoma* spermary or *Ascaris*. As previously seen during the study of mitosis, chromosomes must be duplicated during interphase before the events of cell division can occur. Use Figure 5.1 to identify the different phases of meiosis.

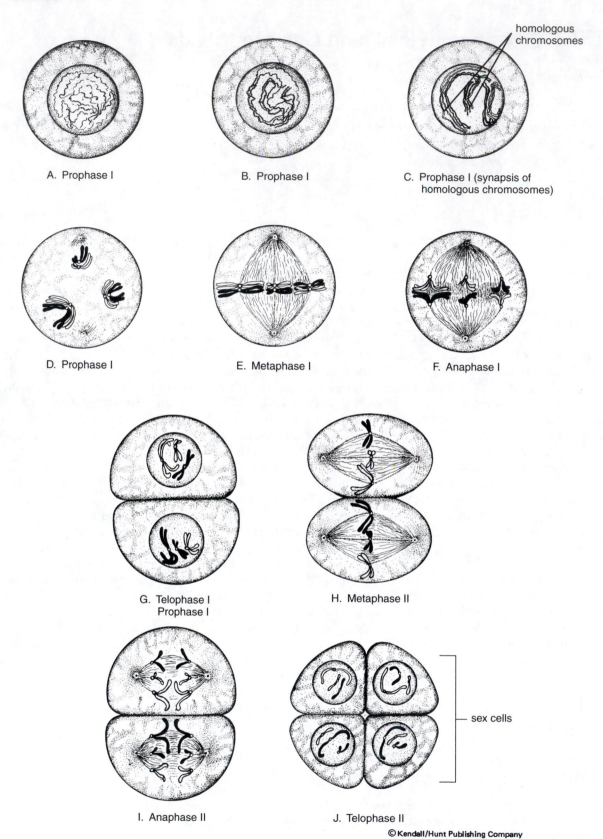

FIGURE 5.1   Meiosis in Animal Cells

First Meiotic Division

Prophase I

The replicated chromosomes (**bivalents**) shorten by coiling and the two separate chromosomes of each **homologous** pair come together. This is termed **synapsis.** The synaptic pair form a group of four **chromatids.** It is in this stage that crossing over and exchange of genetic pieces occurs.

Metaphase I

The nuclear membrane disrupts, groups of four chromatids (**tetrads**) line up along the equatorial plane, and members of each tetrad separate slightly to form two pairs of chromatids.

Anaphase I

Each chromosome, still consisting of its two chromatids, migrate toward one of the two centrioles. This process is also known as **karyokinesis** which means the division of the nucleus.

Telophase I

The cell now divides into two separate cells with each containing a set of chromosomes. This process is known as **cytokinesis.** The nuclear membranes reform marking the end of division I. The chromosomes still consist of two chromatids.

Second Meiotic Division

Prophase II

This division may occur without a preceding interphase as a direct continuation of telophase I.

Metaphase II

The chromosomes, still consisting of two chromatids, line up along the equatorial plane within each cell boundary. The **centromeres** divide, allowing separation of the chromatids and their migration toward opposite poles.

Anaphase II

The chromatids, now chromosomes, began migration toward each of the two centrioles within the boundary of each cell.

Telophase II

Nuclear membranes form around each group of chromosomes. There is cytoplasmic separation (*cytokinesis*) into four distinct cells. Note that each nucleus is now haploid, containing only one daughter chromosome from each tetrad.

The meiotic stages are followed by completion of the gametic process, namely the production of matured sperms or eggs.

Questions

1. What is the purpose of meiosis?

2. How do mitosis and meiosis differ in terms of genetic results?

3. How is meiosis related to sexual reproduction?

4. How is meiosis related to genetic variation and to evolution?

5. Explain how asexual and sexual reproduction differ.

EXERCISE 6   Fertilization and Embryology

Learning Objectives

✔ Identify the various stages of development in the starfish egg.

✔ Identify the three germ layers of the embryo and the types of tissues that develop from them.

✔ Distinguish between protostomes and deuterostomes.

Embryology is a specialized branch of biology that studies the development and growth of the **embryo**. The life cycle of most sexually reproducing organisms consists of several distinct phases. Because sea star development after fertilization is very similar to mammalian development, it will be used for examination of a few conspicuous stages of embryonic specialization.

Gamete Production

The first stage, *gametogenesis*, produces gametes that are haploid from diploid cells. Usually, the gametes of the female (eggs) are larger than the gametes of the male (sperm). Obtain a prepared slide of starfish cleavage. Observe under low and high power. Find an unfertilized egg. It can be recognized by its large nucleus and prominent nucleolus.

Developmental Process

Fertilized Egg or Zygote

After the gametes have been formed, they can *fuse* (*syngamy*) into one new cell called a **zygote**. This fusion of gametes is also known as **fertilization**. Fertilization of the egg initiates the development of the embryo. Since each gamete was haploid (one set of chromosomes), the diploid state (two sets of chromosomes) is restored to the zygote. After the sperm head penetrates the egg cytoplasm, a **fertilization membrane** extrudes from under an outer jelly layer to prevent other sperm from entering. The fertilized egg appears dark when stained due to the fertilization membrane.

Cleavage

Embryogenesis encompasses most of the developmental events that the embryo experiences. From this time on, most higher animal groups progress through very similar stages, but the specific details may vary (Figure 6.1 and Plate 6.1).

The zygote must divide repeatedly to become a mass of cells that is integrated structurally and functionally; that is, become multicellular. The egg is very large in most organisms compared to the size of the normal body cell or **somatic** cell. Because of its size, the surface area of the cell is too small to allow sufficient movement of gases, nutrients, and waste through the cell membrane. Therefore, the zygote must reduce the cell size so that these molecules can traverse the cell membrane in sufficient quantities to keep the cell alive. This change in size is brought about by **cleavage**. The early cells of cleavage, **blastomeres**, progressively get smaller each time they divide.

Two-Celled Stage

The zygote is divided into equal meridional halves by a **cleavage furrow** which passes through the animal and vegetal pole. These cells are called blastomeres.

Four-Celled Stage

The first two cells then divide into equal meridional halves, but at right angles to the first division. The result is four equal blastomeres. Subsequent divisions, all at different angles to the preceding division result in an eight-celled stage; 16-celled stage (morula stage in mammals); 32-celled stage, etc.

Blastula

Obtain a prepared slide of the starfish blastula and gastrula. Cleavage results in a single-layered sphere of about 1,000 cells arranged around a central cavity, called the **blastocoel**. The blastula is the approximate size of the original egg due to repeated cleavage of the blastomeres while becoming progressively smaller and more numerous.

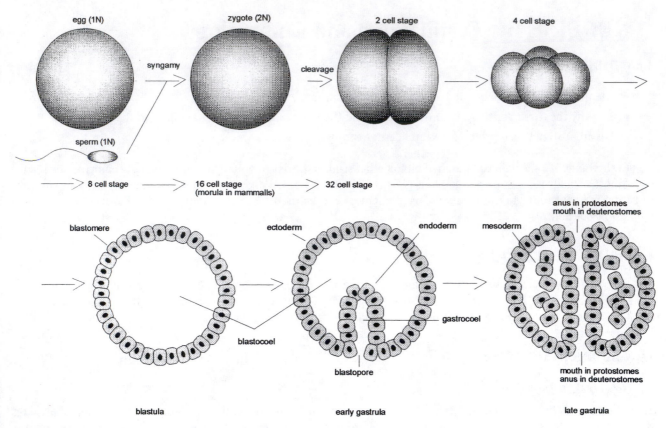

FIGURE 6.1 Developmental sequence of an embryo.

Gastrula

The gastrula is formed from the blastula as a result of the cells in the vegetal pole invaginating into the blastocoel. This is a change from a single-layered to a two-layered embryo by differential growth and migration of cells, a process called **gastrulation**. Gastrulation is the first step in the formation of the **primary germ layers-ectoderm, mesoderm, and endoderm**. As the inner layer forms it gradually crowds out the old blastocoel and forms a new cavity, called the **gastrocoel**. The external opening, as a result of the migration of cells inward, is called the **blastopore**.

Germ Layer Derivatives

Three primary germ layers differentiate in the gastrula and, in turn, form all other tissues in the animal. The *ectoderm* is the outermost layer and, in chordates, it forms the nervous system; the enamel of the teeth; the lens of the eye; outer epidermal structures of the body; and the lining of the mouth, nostrils, and anus. The *endoderm* is the innermost layer and it forms most of the liver and pancreas; thyroid, parathyroid, and thymus glands; and the lining of the following structures: digestive tract, gills, lungs, pharynx, urinary bladder, gall bladder, and urethra. The *mesoderm* is the middle layer and it forms the muscular system, kidneys, loose and dense connective tissues, bone, cartilage, dentine of teeth, dermis of the skin, and reproductive organs.

The digestive tract of animals results from an indentation in the blastula. The initial opening to the digestive tract is called the blastopore. In those animals with an **incomplete gut tract** (cnidarians and flatworms), the blastopore becomes the mouth. Animals with a **complete** digestive tract form an anus as well as a mouth. In **protostomes** (annelids, arthropods, and molluscs), the blastopore forms the mouth while the anus forms at the other end of the digestive tract as an additional opening. In **deuterostomes** (echinoderms, hemichordates, and chordates), the blastopore forms the anus with the mouth forming at the other end of the gut tract.

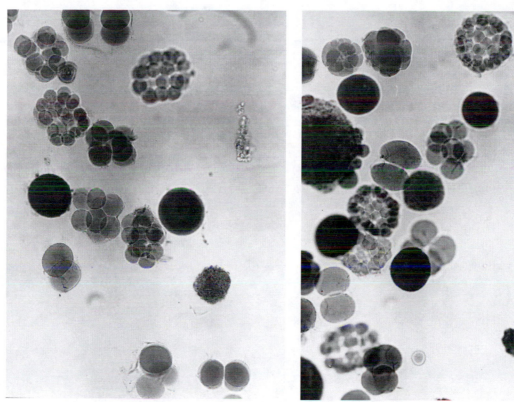

A. Early and late cleavage

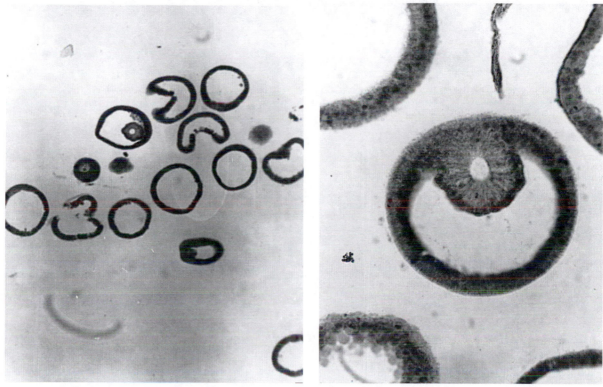

B. Blastula and Gastiula stages

PLATE 6.1   Embryology.

Questions

1. How does the chromosome number of the blastomere compare with that of the zygote?

2. What are the differences between a blastula and a gastrula?

3. What tissues will develop from each of the germ layers or tissues?

4. Why is cleavage necessary?

5. Distinguish between protostome and deuterostome development.

Unit III
A Survey of Invertebrate Animal Phyla

EXERCISE 7   Classification

Learning Objectives

✔ Explain why taxonomy is important to the biological sciences.

✔ Describe how organisms are classified into the major taxonomic groups.

✔ Recognize distinctive characteristics of major animal phyla.

✔ Distinguish between invertebrate and vertebrate animals.

Introduction

Classification, taxonomy or systematics, has many purposes. First, it enables man to identify those animals with which he works to others as well as to himself. Secondly, their classification reflects varying degrees of similarity as well as an evolutionary relationship among those animals which are grouped together. In short, systematics provides for a logical grouping of animals that are similar in body structure.

A Swedish botanist, Carolus Linnaeus, is credited with devising an international system of naming plants and animals scientifically. His system employed **binomial nomenclature** to name animals and plants. Consequently, each organism has a scientific name of two parts. The first part of the name is the **genus** which is followed by the **species** name. The genus begins with a capital letter while the species begins with a small letter. All scientific names are given in Latin. The scientific name is either italicized or is underlined.

The present classification system is a hierarchal system in which the individual organism is linked to others of the same kind within a group or **taxon** (Pl; taxa). Each group encompasses closely related taxa and progressively becomes more inclusive or exclusive dependent upon its direction. Beginning at the most inclusive taxon and progressing with more specificity the taxa are: Kingdom, Phylum, Class, Order, Family, Genus and Species. Below is a summary of the major phylogenetic categories of animals with some of their characteristics. For sake of convenience, several taxa have been omitted.

Kingdom Protista: Unicellular animals

Subkingdom Protozoa

Phylum Sarcomastigophora: Locomotion by flagella, pseudopodia, or both; one nucleus; sexual reproduction by syngamy.

Subphylum Mastigophora (Flagellates): One or more flagella; heterotrophic, autotrophic, or both.

Class Phytomastigophorea (Phytoflagellates): Chloroplast; autotrophic and heterotrophic; *Euglena, Volvox, Peranema, Chilomonas.*

Class Zoomastigophorea (Zooflagellates): No chloroplast; one to several flagella; heterotrophic; *Trichonympha, Leishmania, Trichomonas, Trypanosoma.*

Subphylum Sarcodina (Amoebas): pseudopodia; some with and some without a shell.

Superclass Rhizopoda (Rhizopods): Lobopodia, filopodia, reticulopoda, or no pseudopodia.

Class Lobosea (Amoebas): Mostly lobopodia for pseudopodia; *Amoeba, Arcella, Diffugia, Entamoeba, Pelomyxa.*

Class Granuloreticulosea (Foraminiferans): Granular or clear reticulopoda; *Globigerina.*

Superclass Actinopoda (Actinopods): Axopodia supported by microtubules; spherical; plantonic.

Class Heliozoea (Heliozoans): Skeletal structures of silica or organic material when present; axopodia radiating on all sides; *Actinophyrs.*

Phylum Apicomplexa (Apicomplexans): Apical complex on anterior end; no means of locomotion except for flagellated reproductive cells in some groups; all parasitic.

Class Sporozoa (Sporozoans): Spores or oocyst; flagella in microgametes; *Gregarina, Plasmodium, Toxoplasma, Babesia.*

Phylum Ciliophora (Ciliates): Cilia; two types of nuclei; asexual reproduction by fission, sexual reproduction by conjugation.

Class Polyhymenophorea: Compound ciliary organelles around cytostome; ciliary organelles for locomotion; *Euplotes, Stentor.*

Class Kinetofragminophorea: No compound ciliary organelles around cytosome; *Didinium, Balantidium.*

Class Oligohymenophorea: Compound ciliary organelles around cytostome, usually inconspicuous; *Paramecium, Vorticella, Tetrahymena.*

Kingdom Animalia

Subkingdom Metazoa: Multicellular animals

Phylum Porifera (Sponges): Cellular level of organization; radial symmetry; many pores; internal skeleton; aquatic.

Class Calcarea (Calcareous Sponges): Internal skeleton of limy spicules; asconoid, syconoid, and leuconoid body forms; *Scypha, Leucosolenia.*

Class Demospongiae (Demosponges): Skeleton of spongin or a siliceous skeleton; leuconoid body form; includes freshwater as well as marine varieties; *Spongilla.*

Class Hexactinellida/Hyalospongiae (Glass Sponges): Siliceous spicules; cup shaped; syconoid body form; *Euplectella.*

Class Sclerospongiae (Coralline Sponges): Skeleton of siliceous spicules and spongin; encasement in calcium carbonate; leuconoid body form; *Astrosclera.*

Phylum Cnidaria (Coelenteratans): Tissue level of organization; radial symmetry; polyp and/or medusa; nematocyst; incomplete digestive tract or cavity; aquatic.

Class Hydrozoa (Hydroids): Solitary or colonial; noncellular mesoglea; medusa with velum; *Hydra, Physalia, Obelia, Gonionemus*

Class Scyphozoa (Jelly fishes): Small to large medusa; mesoglea cellular; polyp minute or lacking; *Aurelia.*

Class Anthozoa: All polyps; no medusa; mesoglea cellular; *Astrangia, Metridium, Renilla.*

Phylum Ctenophora (Comb jellies): Tissue level of organization; all marine, biradial symmetry; comb rows (ctenes) present; no polymorphism; tentacles present or absent.

Class Tentaculata: Tentacles present; *Pleurobrachia, Velamen, Ctenoplana.*

Class Nuda: Tentacles absent; body cylindrical; *Beroe.*

Phylum Platyhelminthes (Flatworms): Organ level of development; triploblastic; bilateral symmetry; incomplete digestive tract; excretion by flame cells (excretory organs); a pair of anterior ganglia or nerve ring; usually monoecious; aquatic or terrestrial.

Class Turbellaria: Free-living individuals; body ribbon-like and ciliated; usually a ventral mouth; no hooks or suckers; *Dugesia, Euplanaria.*

Class Trematoda (Flukes): All parasites; body flat with ventral suckers, hooks, or both; *Fasciola, Clonorchis, Schistosoma.*

Class Cestoidea (Tapeworms): All parasites; body long consisting of a scolex and detachable proglottids; *Dipylidium, Diphyllobothrium, Echinococcus.*

Class Monogenea (Monogeneans): Body covering with no cilia; posterior holdfast is hooks, suckers, or both; all parasitic, usually on skin or gills of fish; *Polystoma.*

Phylum Nematoda (Roundworms): Complete digestive tract; triploblastic; small or minute bodies, usually slender; Pseudocoelum bilateral symmetry; free-living in soil or water, or parasitic; anterior nerve ring; simple organs; dioecious.

Class Phasmidea (Ascaris, hookworm, pinworm): A pair of small sensory pouches (phasmids) near the posterior tip of the body; a second pair of sensory organs (amphids), poorly developed, at the anterior end of the body; excretory system with one or two lateral canals; both free-living and parasitic forms; *Ascaris, Necator, Enterobius.*

Class Aphasmidea (trichina worm): Phasmids lacking, but amphids well developed; excretory system with one or more renette or glandular cells; mostly free-living, but some parasites; *Trichinella.*

Phylum Rotifer (Rotifers): Ciliated corona around mouth; mastax (muscular pharynx); parthenogenesis common; body organs are eutelic.

Class Seisonidae: Vestigial corona; body elongated; no vitellaria, two ovaries; marine; males and females of equal size and shape; *Seison.*

Class Bdelloidea: Corona normally with two trochal discs; anterior end can retract; two germovitellaria; parthenogenetic; swimming or creeping forms; males unknown; *Rotaria, Philodina.*

Class Monogononta: Males reduced in size; single germovitellarium; Swimming or sessile forms; *Epiphanes.*

Phylum Gastrotricha (Gastrotrichs): Body with bristles, spines, or scales; ventral surface ciliated; head may be lobed or ciliated and tail may be forked; *Chaetonothus.*

Phylum Kinorhyncha (Kinorhynchs): Body with 13 segments, each with spines, but no cilia; retractile head with circle of spines; small retractile proboscis; *Echinoderes, Kinorhynchus.*

Phylum Nematomorpha (Horsehair Worms): Cylindrical body; rounded anterior end, posterior end rounded with two or three caudal lobes; *Gordius.*

Phylum Acanthocephala (Spiny-headed Worms): Invertible proboscis with rows of recurred spines which serves as an attachment organ; all endoparasites of vertebrates; *Macracanthorhynchus.*

Phylum Entoprocta (Entoprocts): Body cup-shaped with a circle of ciliated tentacles; body with one or several stalks attached to the bottom by an attachment pad; *Loxosoma, Urnatella.*

Phylum Mollusca (Molluscs): Bilateral symmetry; triploblastic; body soft, covered by mantle that secretes limy shell; open circulatory system; respiration by gills (excretory organs); excretion by kidneys (excretory organs); complete digestive tract; coelom reduced; ventral muscular foot for locomotion; nervous system typically of three pairs of ganglia; usually dioecious; aquatic or terrestrial.

Class Polyplacophora (Chitons): Shell of eight dorsal plates *Chiton.*

Class Pelecypoda/Bivalvia (Bivalves): Ventral foot hatchet-shaped; body enclosed by two lateral shells; *Unio, Venus, Anodonta.*

Class Cephalopoda (Squids and Octopus): A large head with eyes; mouth surrounded by eight, ten, or more tentacles; *Loligo, Octopus, Nautilus, Sepia.*

Class Gastropoda (Snails and Slugs): Body usually with a shell that is coiled, reduced, or absent; a distinct head; commonly with eyes and tentacles, is present; large flat foot for locomotion; *Helix, Busycon, Physa.*

Class Scaphopoda (Tooth Shells): Small mollusc with a tubular shell which is open at both ends; gills absent; *Dentalium.*

Class Monoplacophora (Monoplacophorans): Single arched shell; flat, broad foot; *Neopilina.*

Class Aplacophora (Solenogasters): No shell, foot, or mantle; poorly developed head; worm-like; *Neomentia.*

Phylum Annelida (Segmented Worms): Body segmented; bilateral symmetry; triploblastic; setae for appendages; well-developed coelom; complete digestive tract; closed circulatory system; respiration by gills or by epidermis; excretion by nephridia, usually one pair per somite; dorsal ganglia with solid, ventral nerve cord; free-living or parasitic; monoecious or dioecious; aquatic or terrestrial.

Class Oligochaeta (Earthworms): Segmentation conspicuous both internally and externally; no head; monoecious; clitellum secretes cocoon; *Lumbricus, Tubiflex.*

Class Polychaeta: A head with tentacles; parapodia; sexes usually separate; *Nereis (Neanthes), Arenicola.*

Class Hirudinea (Leeches): Flat body with anterior and posterior suckers; no appendages; *Hirudo, Placobdella.*

Phylum Tardigrada (Water Bears): Nonchitinous exoskeleton; body unsegmented; four pairs of short, unjointed legs with four to eight claws on trunk; *Macrobiotus, Echiniscus.*

Phylum Onychophora (Velvet Worms): No external segmentation; paired appendages; outer body covering soft and velvety; antennae on head; *Opisthopatus, Peripatus.*

Phylum Arthropoda (Joint-Foot Animals): Body segmented; bilateral symmetry; triploblastic; appendages jointed; three body sections-head, thorax, and abdomen; exoskeleton; complete digestive tract; open circulatory system with a dorsal heart; coelom reduced; respiration by tracheae, gills; or book lungs; excretion by green glands or Malpighian tubules; dorsal ganglia with paired ventral nerve cord; simple and compound eyes; usually dioecious; aquatic or terrestrial.

Subphylum Chelicerata (Chelicerates): Six pairs of appendages: first pair chelicera (claw-like), second pair pedipalps, four pair of legs; no antennae; body composed of two regions, cephalothorax and abdomen which is usually nonsegmented.

Class Merostomata (Horseshoe Crab): Aquatic with book gills; sharp spine-like telson; two compound eyes and two simple eyes; *Limulus.*

Class Arachnida (Spiders, Scorpions, Mites, and Ticks): Terrestrial with book lungs, tracheae, or both; simple eyes only; *Argiope, Loxosceles, Trombicula, Demodex, Dermacentor, Centuroides.*

Class Pycnogonida (Sea Spiders): Small abdomen; four to six pairs of walking legs; no special excretory or respiratory structures; mouth on long proboscis; *Pycnogonum.*

Subphylum Crustacea (Crustaceans): Body composed of two sections, cephalothorax and abdomen; two pairs of antennae; five pairs of walking legs; respiratory organs are gills; excretory organs are green glands; three pairs of jaws.

Class Cirripedia (Barnacles): Calcium Carbonate shell enclose animal; sessile; head reduced; abdomen absent; legs with long jointed cirri with setae; hermaphroditic; *Balanus, Lepas.*

Class Malacostraca (Lobsters, Crayfish, Crabs): Body with abdomen and cephalothorax that projects forward to form a carapace; *Cambarus, Homarus, Uca.*

Class Branchiopoda (Fairy and Brine Shrimp, Water Fleas): Some with and some without a carapace; reduced first antennae and second maxillae; legs flatten and leaf-like and functioning as gills and for filter feeding; *Daphnia.*

Class Ostracoda (Ostracods): Body encased in a bivalved carapace; thoracic appendages two or none; *Asterope, Cypris, Cytherella.*

Class Copepoda (Copepods): No carapace; single eye; one pair of uniramous maxillipeds; four pairs of thoracic appendages; *Calanus.*

Subphylum Uniramia (Insects and Myriapods): Appendages uniramous (one branch); one pair of antennae; one or two pair maxillae, and one pair mandibles

Class Chilopoda (Centipedes): A long, flattened body of many somites with one pair of appendages; *Lithobius.*

Class Diplopoda (Millipedes): A long, cylindrical body of many somites with two pairs of appendages per abdominal somite; *Spirobius.*

Class Insecta (Insects): Head, thorax and abdomen distinct; usually two pairs of wings; three pairs of legs; one pair of antennae; salivary glands with digestive system; respiration generally by tracheae; excretion by malpighian tubules; dioecious; *Romalea.*

Class Pauropoda (Pauropods): Small bodies with 12 segments; 9-10 pairs of legs; body cylindrical with double segments; no eyes; *Pauropus.*

Class Symphyla (Garden Centipedes): Resemble centipedes; 15-22 segments; 10-12 pairs of legs usually; filiform antennae; *Scutigerella.*

Phylum Bryozoa/Ectoprocta (Moss Animals): Colonial with each member (zooid) encased in a chamber (zoecium); zoecium or exoskeleton may be limy, gelatinous, contain sand, or chitinous; zoecium varies in shape; *Bugula* (marine), *Plumatella* (freshwater).

Phylum Brachiopoda (Lamp Shells): Attached; most forms stalked with dorsal and ventral calcareous plates; *Lingula; Glottidia, Terebratella.*

Phylum Echinodermata (Echinoderms): Radial symmetry in adults; bilateral in larvae; triploblastic; no segmentation; water vascular system; mesodermal endoskeleton; circulatory system reduced; respiration by dermal branchiae; nervous system with circumoral ring and radial nerves; dioecious; aquatic.

Subphylum Asterozoa: Star-shaped with pentamerous radial symmetry; body with separate rays projecting from a central disc; unattached as adults.

Class Asteroidea (Starfish): Body star-shaped with 5 to 50 rays indistinct from the disc; open ambulacral groove with two to four rows of tube feet; tube feet with suckers and used for locomotion; flexible endoskeleton; pedicellariae; *Asterias, Acanthaster.*

Class Ophiuroidea (Brittle Stars and Basket Stars): Body star-shaped with five or more rays distinct from the disc; ambulacral grooves closed and covered by ossicles; tube feet without suckers and not used for locomotion; pedicellariae absent; *Ophiothrix, Astrophyton.*

Subphylum Echinozoa: Globoid, discoid, or cylindrical; no arms.

Class Echinoidea (Sea Urchins and Sand Dollars). Hemispherical body with moveable spines; no separate rays; ambulacral grooves covered with ossicles; tube feet with suckers; pedicellariae present; shell of closely fitting plates; *Arbacis* (sea urchin), *Encope* (sand dollar).

Class Holothuroidea (Sea Cucumbers): Cucumber-shaped body with no rays and a leathery body; no spines; ambulacral groove closed; tube feet with suckers; pedicellariae absent; *Cucumaria.*

Subphylum Crinozoa/Pelmatozoa: Radially symmetrical; round or cup-shaped theca and arms; attached by stem.

Class Crinoidea (Sea Lilies and Feather Stars): Body stalked with five rays that branch; madreporite, spines, and pedicellariae absent; ambulacral groove and tentacle-like tube feet present; *Antedon, Nemaster.*

Phylum Chaetognatha (Arrowworms): Worm-like; soft bodied; hood over head; bristles around mouth; *Sagitta.*

Phylum Hemichordata (Hemichordates): Worm-like; epidermal nervous system; pharyngeal gills; body with proboscis; collar, and trunk.

Class Enteropneusta (Acorn Worms): Tongue-like proboscis; long trunk; short collar; *Saccoglossus, Balanoglossus.*

Class Pterobranchia (Pterobranchs): Body with proboscis, collar, and trunk; one pair gill slits; two or more arms; *Cephalodiscus.*

Phylum Chordata (Chordates): Notochord; paired gill slits; dorsal, tubular nerve cord, post-anal tail.

Group Acrania (Prochordates)

Subphylum Urochordata (Tunicates): Only the free swimming larvae has notochord and nerve cord; adults with tunic and sessile.

Class Ascidiacea (Sea Squirts): Sessile as adults; colonial or solitary; *Botryllus.*

Class Larvacea (Larvaceans): Adults retain notochord and tail; no tunic; epidermis produces gel-like covering for body.

Class Thaliacea (Thaliaceans): Adults without tails, barrel shaped; pelagic; incurrent and excurrent siphons at opposite ends; *Doliolum.*

Subphylum Cephalochordata (Lancelets): Fish-like; no scales; notochord and nerve cord along entire length of body; many gill slits; *Branchiostomata* (*Amphioxus*).

Group Craniata (Vertebrates)

Subphylum Vertebrata (Vertebrates): Body covering scales, skin, feathers or hair; internal skeleton; skeletal muscles for locomotion; digestive tract complete with ventral mouth; with tongue and teeth; circulatory system closed; heart two, three, or four chambered; poikilothermic or homeothermic; respiration by gills or lungs; excretion by paired kidneys; endocrine glands; brain with 10 to 12 cranial nerves; dioecious.

Superclass Agnatha/Cyclostomata (Jawless Vertebrates): No true jaws or appendages.

Class Myxini (Hagfish): Terminal mouth with four pairs of tentacles; degenerate eyes; slime glands; one pair of common gill openings; no dermal bone; nasal sac with duct to pharynx; *Myxine, Bdellostoma.*

Class Cephalaspidomorphi/Petromyzontes (Lampreys): Suctorial mouth with horny teeth; rasping tongue; separate gill openings; moderately developed eyes; nasal sac not connected to pharynx; *Petromyson, Lampetra.*

Superclass Gnathostomata (Jawed Vertebrates): With jaws and usually have paired appendages.

Class Chondrichthyes (Cartilaginous Fish—Sharks, Skates, Rays, and Chimaeras): Cartilaginous skeleton; heterocercal tail; five to seven gills with separate openings; no operculum; no swim bladder; poikilothermic (ectothermic); viviparous, oviparous, or ovoviviparous; *Squalus, Raja, Chimaera.*

Class Osteichthyes (fishes): Bony skeleton; usually homocercal tail; jaws; scales; terminal mouth; gills covered by an operculum; two chambered heart; poikilothermic; oviparous or ovoviviparous; *Salmo, Perco, Latimeria.*

Class Amphibia (Amphibians): Moist glandular skin; no scales; two pairs of limbs for walking or swimming; nostrils (two) connected to mouth cavity; bony skeleton; three-chambered heart; respiration by skin, gills, lungs or mouth; poikilothermic; oviparous; *Plethodom, Rana, Necturus, Bufo, Xenopus.*

Class Reptilia (Reptiles): Dry, scaly skin; two pairs of limbs; five toes with claws; heart with incompletely divided ventricles or completely divided ventricles; lungs; poikilothermic; ovoviviparous or oviparous; *Gekko, Crotalus, Alligator.*

Class Aves (Birds): Feathers; forelimbs modified as wings; hindlimbs for perching, walking, or swimming; beak or horny bill; four-chambered heart with right aortic loop; lungs; no urinary bladder; homeothermic; oviparous; *Gavia* (loon), *Pelecanus* (pelican), *Ardea* (heron), *Anser* (geese), *Anas* (ducks).

Class Mammalia (Mammals): Hair; mammary glands; teeth of several types; seven neck vertebrae; pinnae; mobile tongue; heart four chambers with left aortic loop; lungs; diaphragm; urinary bladder; brain highly developed; homeothermic; viviparous; *Sorix* (shrews), *Tadarida* (bat), *Lepus* (hare), *Canis* (dog), *Felis* (cat).

Laboratory Exercise

Obtain a box of ten little "animals" from the demonstration desk. You have 15 minutes to classify these "animals" according to their characteristics. Try to make a dichotomous key. A dichotomous key is made by pairing characteristics. Separate these "animals" into two groups. Try to determine these two groups before going on to the next level. After you have determined these two groups, proceed to the next level until you have completed your key.

The key below will aid your classification scheme. Notice that each statement has a number and is followed by a number in parenthesis. If the statement agrees to your finding, then go to the next numeral in sequence and so on; however, if it does not agree, go to the one in the parenthesis. This classification scheme is called a dichotomous key.

1(7)	one-celled animals	Kingdom Protista
2(6)	organelles for locomotion	
3(4)	pseudopodia	Subphylum Sarcodina
4(5)	cilia	Subphylum Ciliophora
5(3)	flagella	Subphylum Zoomastigophora
6(2)	no organelles for locomotion	Class Sporozoa
7(1)	multicellular animals	Kingdom Animalia

Questions

1. Why do biologists classify organisms?

2. What levels of classification are used to classify animals from species to kingdom?

3. Does this classification refer to individuals, as our own names do, or to groups of individuals?

4. What is a biological species?

5. Could organisms be classified by criteria other than by blood relationships? If so, how?

EXERCISE 8   Life Processes

Learning Objectives

✔ Name the three general attributes of all living organisms.

✔ List the specific processes necessary to maintain life.

✔ Describe the different levels of organization that living organisms possess.

Characteristics of Life

Before beginning the study of the animal kingdom, it is important to consider some of the characteristics of organisms. Life is expressed in countless forms, from the simplest single celled organisms to the most complex of all animals, man. But all animal forms can be characterized by several attributes including; 1) a constant state of activity, 2) a definite life cycle or span and 3) a definite form and size.

Each organism progresses through its life cycle from the first state of "being" through growth stages, maturation and ultimately death. The life span varies with each organism, ranging from days to decades. During the growth and development phase, body size increases and reaches a plateau stabilizing in a definite body size. The progression to full maturity is accomplished through the chemical activity carried on within the body. Whether physically moving about, permanently attached or even in a hibernated state, animals possess a constant state of internal activity. The energy derived from that chemical activity drives the processes of life. Once full maturity has been reached, life processes begin to slow down with aging. Eventually the active processes will cease to function, ending the life cycle at death. Progressing through the life stages can be detailed in specific universal processes.

Life Processes

Life, at the simplest level of existence, is also characterized by a similar chemical state. All life on this planet has been identified as being composed of carbon-containing compounds. This organic composition is seen in the chemical make-up of all living cells; namely, the organic compounds of carbohydrates, proteins, lipids and nucleic acids. All of these chemicals have a basis of carbon atoms along with other critical elements. These building blocks are assembled into functional and structural units of cellular organization. Organizing and disassembling chemicals results in cellular activity (metabolism) and an increase in size or growth.

Ingestion is the taking in of food. This is a major difference between plants and animals. While plants can photosynthesize and produce their own food, animals must search for and ingest food material to maintain their existence. Digestion is closely associated to ingestion but involves the chemical and physical breakdown of food to make it a suitable size to enter cells. Any unused, undigested or waste material must be removed in the process of egestion by single celled organisms or excretion in more complex body forms.

Assimilation takes the digested food elements and converts them into useful cell or body parts. Cellular secretions are produced by cells or organisms and placed outside the surface for various purposes. Digestive organs secrete enzymes to break down food while glandular structures secrete hormones to regulate activity. Any living unit must also be quick to react to the environment; both internal and external. Cells possess irritability; a response to a stimulus. This can be seen in the light response of single celled, swimming organism up to the highly reactive pain reflex of mammals.

A constant state of internal activity identifies living cells but external activity is a common process in animals. The ability to move about (locomotion) identifies many different groups of animals, such as flying, running, and swimming.

All life comes from preexisting life; hence "making more of the same" is accomplished through reproduction. By simply breaking into smaller pieces (fragmenting) or growing a small form off a large body (budding), organisms can produce new offspring by asexual reproduction. When uniting two different cells or bodies in sexual fertilization, new generations are produced with a variety of traits. Either method of reproduction affords uninterrupted life and life processes. Continued life on this planet is dependent upon preserving the conditions in which living organisms can maintain their existence, accomplish their activities and replace themselves.

In Exercise 3, various levels of hierarchy were introduced. In higher organisms, different molecules form organelles or cell parts. The unified single cell carries on life processes and acts independently. Tissues, organs and systems in complex animals accomplish the same life activities except in a more complicated and intricate manner. As each new level of organisms is addressed or each new system is introduced, notice the various methods that are used to accomplish the universal life processes.

EXERCISE 9   Kingdom Protista

Subkingdom Protozoa

Unicellular Level of Organization

Learning Objectives

✔ Describe the organelles and methods of movement that are unique to the various groups of protists.

✔ Identify representative members of the different groups of protista with the microscope.

✔ Explain the ecological role these organisms play in the environment.

✔ Explain how the different varieties of protists affect humans medically and economically.

Introduction

Protistans are microscopic one-celled organisms. They are among the simplest of creatures yet are often complex in function. They carry on all the metabolic activities delegated to various tissues, organs, and systems found in higher organisms except in a single cell form. They occur in all kinds of habitats; e.g., the sea, fresh water and even in the soil. Certain protistans are ectoparasites or endoparasites while others have predominant plant or animal characteristics. Presently the protistans include two subkingdoms: the plant-like algae and the more animal-like protozoans. While the algae are more traditionally introduced in Botany courses, this lab exercise will focus on the protozoan examples.

The diverse group of protozoan are classified by many features including: types of nucleus, reproductive processes, methods of locomotion, nutrition, and cellular coverings. The separation into phyla is generally by devices which include no active means of locomotion (spore formers), pseudopodia, flagella, cilia.

Phylum Sarcomastigophora

The sarcomastigophorans are animal-like protozoans recognized by the ability to move by flowing cytoplasmic extensions, or **pseudopods**, and tail-like **flagella**. Because examples may possess both locomotive devices at some time in their life cycle, this large phylum includes organisms with one or both methods. Two of the subphyla, Sarcodina and Mastigophora, are discussed here.

Subphylum Sarcodina

Example: *Amoeba proteus*

The freshwater amoebae are usually found on the bottom of ditches, ponds, or slow-moving streams, and on plant material in water. They move with pseudopods during their active stage and may become dormant in a cyst form during adverse conditions such as dry periods. Freshwater amoeba are difficult to maintain in laboratories; consequently, they are obtained from biological supply houses.

Slide Preparation of Living Protozoan Specimens

On the demonstration table are cultures of various live protozoans. From a jar containing live *Amoeba* prepare a wet mount slide by the following "hanging drop" method. Take a toothpick and obtain a small amount of vaseline. Smear the vaseline around the edge of the cover glass to create a depression that will hold a drop or two of water. Using a pipette, obtain a drop of *Amoeba* culture from the debris on the bottom of the jar. Drop the culture in the vaseline square on the cover glass. Gently lower the slide down on the cover glass containing the culture and turn the slide over. The cover glass and culture should now be on top of the slide and can be examined under low power. The vaseline serves a double purpose. It prevents drying out and protects the *Amoeba* from crushing.

Locate the cells and allow the *Amoeba* to adjust to the new environment. Watch for the movement of **pseudopodia**. These are finger-like projections extending from the main mass of cell structure. The liquid matrix of the cell is

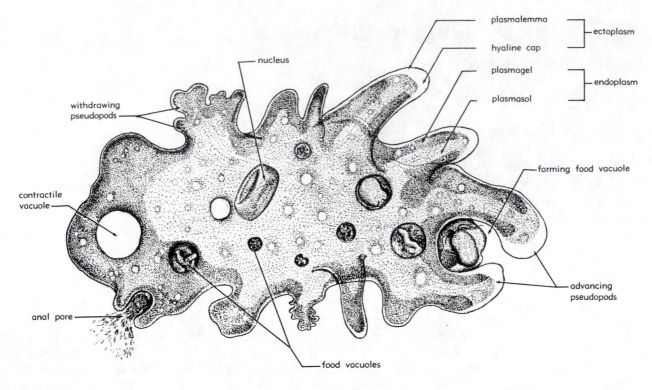

nucleus

plasmalemma — ectoplasm

hyaline cap

plasmagel — endoplasm

plasmasol

withdrawing pseudopods

forming food vacuole

contractile vacuole

advancing pseudopods

anal pore

food vacuoles

FIGURE 9.1 Amoeba

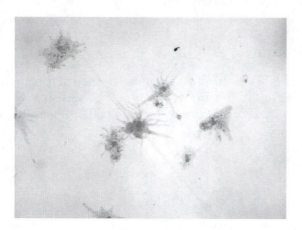

PLATE 9.1 Amoeba

marked by physical streaming. The extension and subsequent flow of cytoplasm results in a type of locomotion called *"amoeboid movement."* There may be marked differences in the appearance of the cytoplasm. The clear **ectoplasm** is generally the area adjacent to the cell membrane while the more granular **endoplasm** fills the cell. As movement occurs, the texture of the cytoplasm interchanges between a firmer **plasmagel** state to the more fluid **plasmasol** state.

The **food vacuoles** are small organelles filled with granular food material and water which have been engulfed by pseudopodia. Intracellular digestion occurs within these fluid-filled vacuoles. The process of digestion is indicated by a decrease in size of food particles and the vacuole. Notice the discharging of undigested wastes (egestion). In this process the food vacuole moves to the cell membrane, then the cytoplasm moves away from it, leaving the waste outside the cell.

The **nucleus** is disc-like in side view, and oblong in edge view. In living wet mount preparations, the nucleus may appear translucent and therefore difficult to see. Locate a clear, round structure about as large as the nucleus and lying close by. Observe the **contractile vacuole** swelling, bursting, and then reforming once again. The contractile vacuole regulates the amount of water in protozoans. During feeding, excess water enters the cell in the food vacuole and must be excreted constantly by this cell organelle.

Check the response of the living amoeba by changing the light. Be patient as their movement is slow. Observe its response to a light tap on the cover glass. If available, introduce a food source, such as *Chilomonas,* into your wet mount and observe the "engulfing" of solid material by pseudopodia.

General Structure

Examine a stained slide of *Amoeba proteus,* the traditional sarcodine (Figure 9.1, Plate 9.1). Note the general outline and structure. Identify the cell components: *cell membrane (plasmalemma), cytoplasm, pseudopodia, contractile vacuole (water vacuole), nucleus, and food vacuoles.* If these parts are difficult to see, vary light intensity with the iris diaphragm. Learn to achieve a three-dimensional effect by constantly moving the fine adjustment.

Other Sarcodines

Examine other live sarcodines on display and make wet mounts of them. Unlike the "naked" amoebas just observed, some sarcodines are covered by shells or tests of various compositions. In the foraminiferans (*Globigerina* sp.), the shells have multiple, perforated chambers and are generally composed of calcium carbonate (limestone). Members of the radiolarian group include exquisite siliceous shelled samples (Plates 9.2, 9.3). Many amoebas are parasitic forms and may be ingested with contaminated foods. One of the familiar human parasites causing amoebic dysentery is *Enatamoeba histolytica.* Using live or prepared specimens, observe these and other sarcodines such as: *Peloymxa carolinensis* (*Chaos chaos), Diffugia lobostoma,* and *Arcella vulgaris* (Figure 9.2).

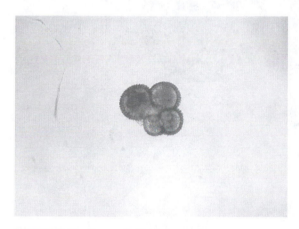

PLATE 9.2   Foraminifera

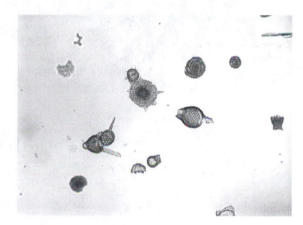

PLATE 9.3   Radiolarians

Figure A. Ameba

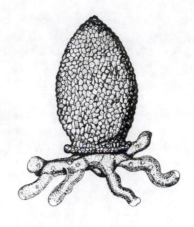

Figure B. Difflugia

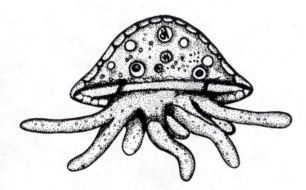

Lateral View

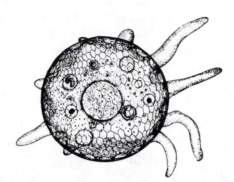

Surface View

Figure C. Arcella

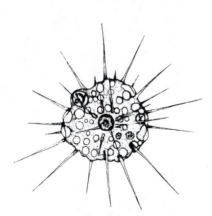

Figure D. Actinophrys

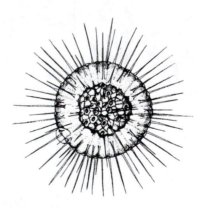

Figure E. Actinosphaerium

(Not drawn to scale)

© Kendall/Hunt Publishing Company

FIGURE 9.2 Various Sarcodines

Questions

1. What color do the live amoebae appear to be?

2. Describe the shape or form of amoebae.

3. Does the amoeba move constantly in one direction? Why?

4. Describe "amoeboid movement."

5. Distinquish between ingestion, digestion, and egestion in the amoeba.

6. What is implied by a high rate of pumping in the contractile vacuole? A slow rate?

7. What species looked as though it was pebbled?

8. What is the geological significance of foraminiferan deposits?

9. In what way do the *Difflugia* resemble *Amoeba?*

Phylum Sarcomastigophora

Subphylum Mastigophora

The mastigophorae are protozoans identified by their locomotive tail-like flagella. Active swimmers, they are grouped into two classes; the phytomastigophorans (phytoflagellates) and the zoomastigophorans (zooflagellates) dependent upon their ability to photosynthesize. Some members of this subphylum include aggregations of flagellated cells called colonies.

Class Phytomastigophorea

Example: *Euglena viridis*

Species of *Euglena* are common in ponds and sluggish streams, sometimes in numbers large enough to produce a reddish-green color or scum on the pond surface. The phytomastigophora or phytoflagellates have one, two, or more whip-like flagella that serve for locomotion. Another main characteristic is the presence of chlorophyll-containing bodies called **chloroplasts**. Hence the phytoflagellates are both plant-like autotrophs capable of producing their own food and animal-like in their ability to swim.

Make a wet mount from the jar of live *Euglena*. Locate the flagellum. The light may have to be adjusted to see this structure. The flagellum is a whip-like organelle located at the anterior end. It lashes out in many directions at a rapid speed. Notice the squirming action called "euglenoid movement." The euglena may also move by a third method known as a "spiral crawl."

Locate the chloroplasts that contain a green pigment called chlorophyll. This pigment makes it possible for euglena to carry on photosynthesis. An organism which can produce its own food is known as an **autotroph**. Due to this fact, many taxonomists consider euglena and phytoflagellates in general to be plants rather than animals. The food that is manufactured is stored in structures called **paramylum** bodies. An organelle associated with food manufacturing is the **stigma** or red eyespot. This organelle is sensitive to light. In times of darkness or periods without light, the euglena ingests food by absorption through the membrane. Organisms which cannot manufacture their own food are considered to be **heterotrophic**. *Euglena* and other phytoflagellates present a taxonomic problem because of their plant and animal-like characteristics. Examine a stained slide of *Euglena* (Figure 9.3, Plate 9.4). Note the general shape and structures identified above.

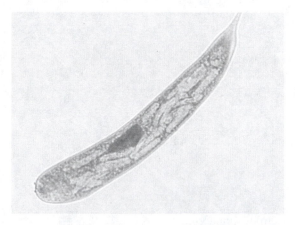

PLATE 9.4   Euglena

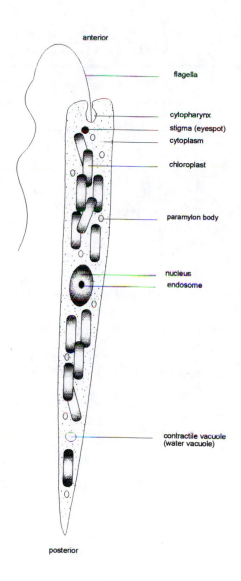

anterior

flagella

cytopharynx
stigma (eyespot)
cytoplasm

chloroplast

paramylon body

nucleus
endosome

contractile vacuole
(water vacuole)

posterior

FIGURE 9.3 Euglena

Questions

1. How does a euglena react to light?

2. Is a euglena a plant? An animal? Give reasons for your choice.

3. What are the paramylum bodies? How are they related to photosynthesis?

4. How does amoeboid movement differ from euglenoid movement?

Class Phytomastigophorea

Example: *Volvox*

Volvox is an unusual colonial flagellate. Each cell in the colony is a biflagellated protozoan. *Volvox* is composed of two kinds of cells that differ in structure and function. The **somatic** (body) cells of the outer wall synthesize food and serve in locomotion; but, eventually all will die. The second type of cells, *sex* cells, serve for sexual reproduction (Figure 9.4, Plate 9.5).

To reproduce, somatic cells lose their flagella and enlarge as spherical **parthenogonidia**. These multiply within the central cavity of the mother colony to become "**daughter colonies**" which escape later to become new colonies. This is a form of asexual reproduction. After several asexual generations, some somatic cells enlarge and become **ova** (*macrogametes* or eggs). Other somatic cells of the same or different colony divide repeatedly to become flat bundles of up to 128 slender **sperm** (*microgametes*). The egg is fertilized by the sperm in the central cavity, resulting in a **zygote**. The zygote can start a new colony in the spring after surviving a hard winter. The old colony of somatic cells dies during the winter.

Volvox clearly represents a high level of protistan complexity with controlled movement by intercellular communicating fibers and specialization of reproduction.

Other Phytoflagellates

Dinoflagellates represent a varied group of phytoflagellates recognized by the presence of a cellulose shell or test. The body has two furrows, one transverse and one longitudinal, each associated with a single flagellum. *Noctiluca* is a familiar marine form capable of bioluminescence. Upon agitation, swarms of *Noctiluca* will cause the waters to "glow." Other dinoflagellates, such as *Gonyaulax* and *Gymnodinium,* are associated with rapid reproductions or "blooms". Large concentrations of these organisms can produce discolored water or "red tides" and kill surrounding fish with accumulated poisons. Observe prepared slides or prepare wet mounts from cultures of other phytoflagellates (Figure 9.4, Plate 9.6) using the procedure described above.

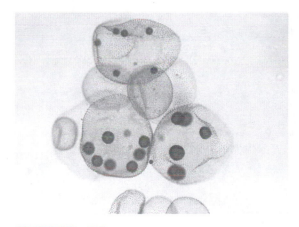

PLATE 9.5   Volvox

PLATE 9.6   Dinoflagellate

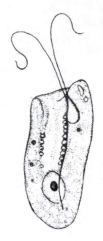

Figure A. Chilomonas

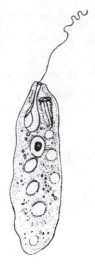

Figure B. Peranema

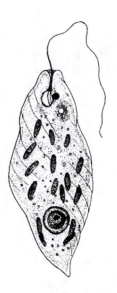

Figure C. Euglena

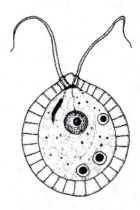

Figure D. Hematococcus

Figure E. Phacus

(Not drawn to scale)

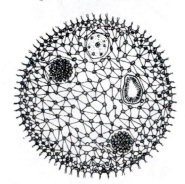

Figure F. Volvox

FIGURE 9.4   Varied Flagellates

Questions

1. Justify the inclusion of *Volvox* within the Kingdom Protista even though it acts like a cellular organism to some degree.

2. How does *Volvox* show cell specialization? Division of labor?

3. What are the advantages and disadvantages of colonial life?

4. What is the medical, monetary and ecological impact of "red tide"?

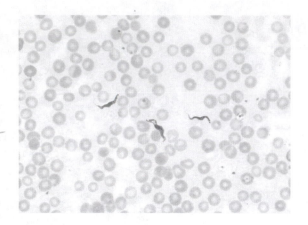

PLATE 9.7   Trypanosoma

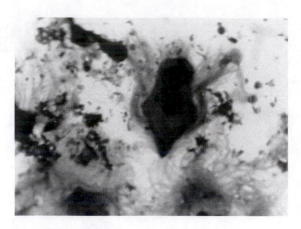

PLATE 9.8   Trichonympha

Class Zoomastigophorea

Example: *Varied zooflagellates*

The zooflagellates lack chloroplasts and paramylum bodies and the ability to photosynthesize. They possess one, two or many flagella and many are internal parasites of higher organisms, including man. Species of *Trypansoma* are blood parasites transmitted by blood-sucking invertebrates. *Trypansoma gambiense* invades the blood through the bite of a tsetse fly causing African sleeping sickness. This debilitating disease results in fever, lethargy, and eventually coma and death. A species of *Trichomonas,* another zooflagellate parasite, is passed through sexual contact and causes vaginitis. Mountain streams, well water and community water ponds may be infected with *Giardia,* which lives in the human intestinal tract and causes nausea, abdominal cramps and diarrhea. The multi-flagellated *Trichonympha,* located in the gut of wood roaches and termites, is responsible for the digestion of plant cellulose. Observe prepared slides of these or other zooflagellates. (Plates 9.7, 9.8).

Phylum Ciliophora

Example: *Paramecium caudatum*

The most amazingly diverse protozoans occur in the ciliates. They include predators, browsers, cruisers, and snake-like searchers. The presence of hair-like **cilia** provide and allow for locomotion and food capture. Ciliates are also identified by the presence of two or more nuclei. When two are present, the larger **macronucleus** is predominately vegetative in function, responsible for metabolic and developmental processes. The smaller **micronucleus** becomes active during sexual conjugation. The outside cell membrane is also reinforced with a thickened **pellicle** in which the cilia are embedded.

Paramecium is one of the ciliates most easily obtained. In nature, the animals are most abundant where the water is foul and ill-smelling such as streams that contain sewage and other decomposing organic material. Prepare a wet mount from a living culture of *Paramecium*. Use a clean glass slide that has a concave depression in its central portion. The paramecia will be concentrated at the bottom of the dish around the particles of bacterial scum upon which they feed. With a pipette, draw up some paramecia and a few bits of scum, and place a drop in the depression of the slide. To this add one drop of yeast cells which have been stained with congo red. To reduce the rapid swimming, a drop of a syrupy methyl cellulose solution or prepared Proto-Slo may be added to the slide. Carefully place a cover glass over the depression and observe under low and high power. Watch as these animals rotate on their long axis, lashing their cilia in a synchronized fashion (Figure 9.5, Plate 9.9).

Note the avoiding action of the paramecium when it strikes an obstacle. Does it turn around, back up, or move around the object? Observe the cilia of the **oral groove** beating. The sweeping action around this depression causes a current of fluid to be swept into the oral groove. This can be easily seen since the numerous stained yeast cells will also be swept into the oral groove. Fluid and food will enter the opening or **cytostome** and move down the **cytopharynx** or gullet and detach as a **food vacuole**. Observe other food vacuoles and determine if digestion has occurred. Congo red is an indicator dye that will gradually turn blue as the yeast particles become more acid. In most animals, a large part of digestion takes place in an acid medium. Does this hold true in the food vacuoles of paramecia?

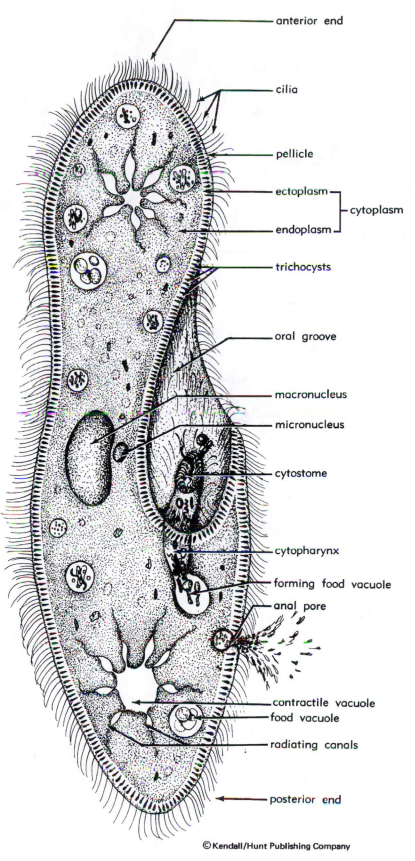

anterior end

cilia

pellicle

ectoplasm

cytoplasm

endoplasm

trichocysts

oral groove

macronucleus

micronucleus

cytostome

cytopharynx

forming food vacuole

anal pore

contractile vacuole

food vacuole

radiating canals

posterior end

FIGURE 9.5 Paramecium

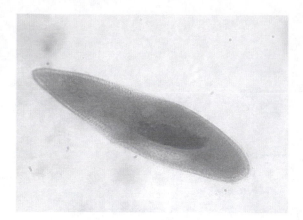

PLATE 9.9 A   Cell Structure

PLATE 9.9 B   Binary Fission

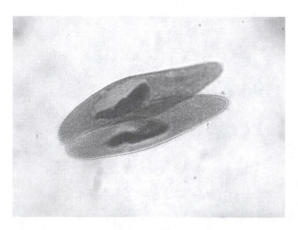

PLATE 9.9 C   Conjugation

To observe the different types of reproduction in the ciliates, examine a prepared slide of paramecia multiplying by *fission* and *conjugation*. Fission is a type of asexual reproduction in which the ciliate divides by mitosis. The cells separate by pulling apart and resulting in new cells end-to-end. Conjugation is a type of sexual reproduction in which nuclear material is exchanged between the micronuclei. At the beginning of this process, the two mating cells lie side by side (Plate 9.9 B, C).

Other Ciliates

A variety of cell shapes occur in the ciliate group. The elongated, torpedo-shape of *Spirostomum* is also recognized by the string of multinuclei along its body axis. Other ciliates may be attached to underwater structures by a stalked segment and sweep the water for food such as species of *Stentor* and *Vorticella*. Although most ciliates are free-living, *Balantidium coli* is an important intestinal parasite of mammals, such as pigs and man. Examine the prepared or live specimens of these or other ciliates (Figure 9.6, Plate 9.10, 9.11, 9.12).

Figure A. Paramecium

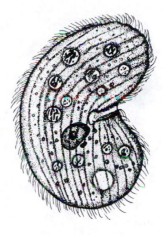

Figure B. Colpoda

Figure C. Stylonichia

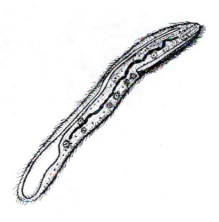

Figure D. Spirostomum

Figure E. Vorticella

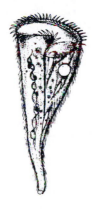

Figure F. Stentor

(Not drawn to scale)

FIGURE 9.6 Varied Ciliates

PLATE 9.10   Spirostomum

PLATE 9.11   Stentor

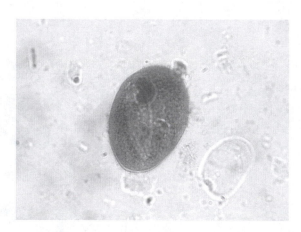

PLATE 9.12   Balantidium

Questions

1. How does the paramecium feed?

2. What changes are there in the size of food vacuoles?

3. Does the acidity of the food vacuoles change?

4. How do paramecia respond to chemicals?

5. Describe the differences between fission and conjugation.

6. Which ciliate is shaped like a trumpet?

7. Which ciliate looks like a ball on a stalk?

8. Did any ciliates possess thick cirri or hairs? If so, which ones?

EXERCISE 10   Phylum Porifera

Primitive Multicellular Sponges

Cellular Level of Organization

Learning Objectives

✔ Describe the basic forms of the sponge body.
✔ Explain the function of the various types of cells.
✔ Identify the skeletal structures and their composition in various sponges.
✔ Explain the relationship of the sponges to the protista and more complex organisms.

Introduction

The Porifera are aquatic, "pore-bearing" sponges, mostly attached to surfaces in marine waters. They exhibit a division of labor among their cells and have reached the cellular level of organization. Not much more than loose aggregations of cells, the sponges are not arranged to form true tissues, organs or organ-systems. They range in size from a few millimeters to nearly 2 meters across. Although sponges are loosely two-layered as adults, there is no equivalent to the developing germ layers of higher organisms. The fertilized zygote of many sponges forms a hollow blastula with flagellated cells lining the interior surface. When the cells reverse position, the flagellated swimming larva is called an **amphiblastula**. The dividing cells develop into the specialized cells of an adult sponge.

The sponge body is generally cylindrical with many small pores or **ostia** through which water passes. Flagellated cells called collar cells or **choanocytes** line inside passages trapping food particles and producing a water current. This type of water sweeping to gather food is called **filter feeding**, a common method found in many aquatic animals. An inside chamber, **spongocoel**, opens to the outside by an upper **osculum**. Giving shape and form to the sponge is an internal skeletal array of protein (**spongin**) fibers and/or spicules of calcium carbonate (limestone) or silicon dioxide (glass). Other cells found in the sponge body include: **pinacocytes** which cover the external surface, **amebocytes** or wandering cells which help distribute food, **archaeocytes** which produce sex cells and **scleroblasts** which secrete the skeletal pieces.

The sponge phylum is subdivided into classes identified by the skeletal composition and body organization (Figure 10.1). Three body types are recognized in the sponges. The **ascon** form is the simplest arrangement of cells. The slender body has many ostia but only one spongocoel, one osculum and calcareous spicules. The **sycon** sponges have a tubular body but the walls are thicker and folded into channels. Water entering the ostia moves into the **inhalent** or incurrent canal and passes through small openings into the **radial** canals. These channels are lined with the collar cells which move the water into the main spongocoel exiting the osculum. The most complex sponge form is the **leucon** type. Clusters of multiple chambers produce a large, nonsymmetrical mass with several oscula.

Asexual reproduction in sponges can include **budding** and **fragmentation**. They are also recognized for their high powers of **regeneration**. Experiments have shown ground up sponge bodies can reassemble cells back into fully formed organisms. Sponges possess both types of gonads resulting in a **monoecious** condition. Sperm cells are released in the water current and fertilize eggs in the chambers of other sponges. The swimming amphiblastula moves to other territories for dispersal.

Class Calcarea (Calcispongiae)

Example: *Scypha or Grantia*

This sponge has an internal skeleton that consists of calcium carbonate spicules. The "body" has many pores opening into one spongocoel and one osculum. Place a preserved sponge in a shallow dish of water and observe with a hand lens or dissecting scope. Identify the ostia and osculum. Obtain a prepared slide of *Scypha* (*Grantia*) and use Figure 10.2 to identify parts of the sponge body in both longitudinal and cross cut sections (Plate 10.1, 10.2). Look at a prepared slide of calcareous spicules (Figure 10.3, Plate 10.3). Be able to recognize the shape and composition of these skeletal units.

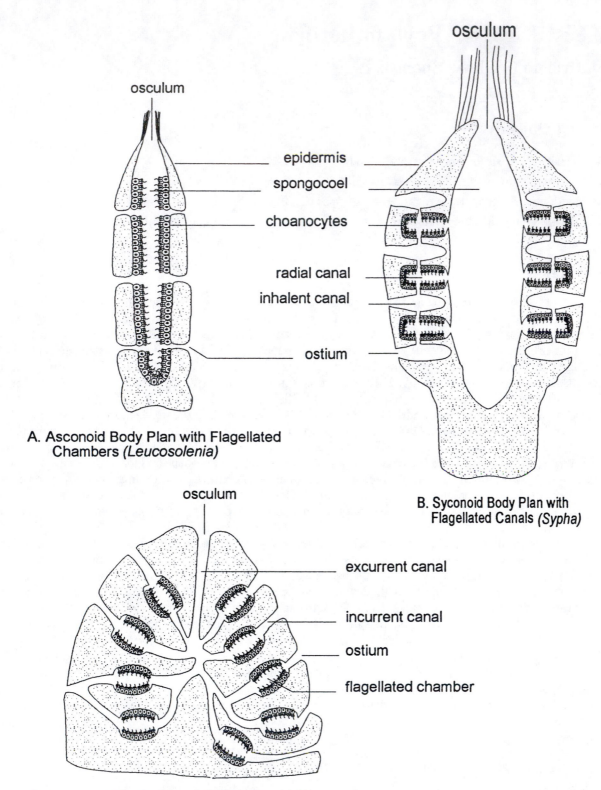

osculum

epidermis
spongocoel
choanocytes

radial canal
inhalent canal

ostium

A. Asconoid Body Plan with Flagellated Chambers *(Leucosolenia)*

osculum

B. Syconoid Body Plan with Flagellated Canals *(Sypha)*

osculum

excurrent canal

incurrent canal

ostium

flagellated chamber

C. Leuconoid Body Plan with Flagellated Chambers *(Euspongia)*

FIGURE 10.1   Sponge Body Plans

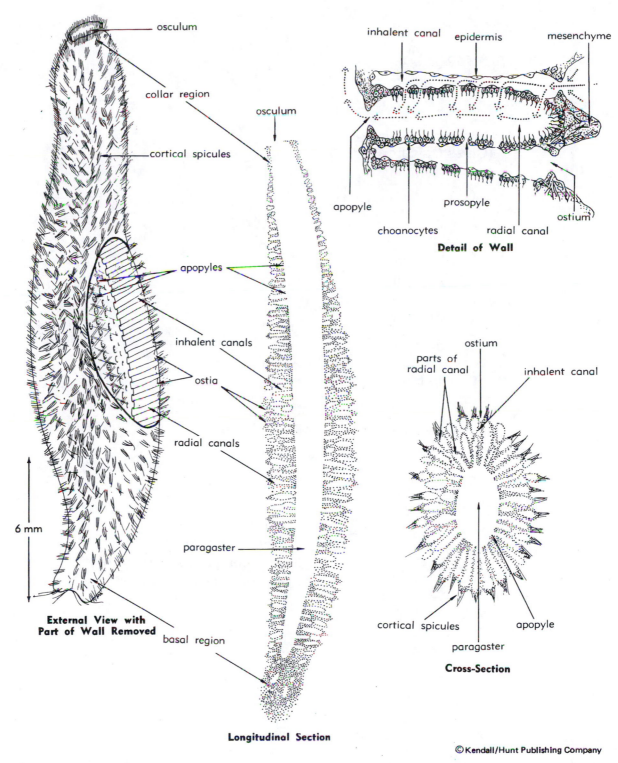

osculum

collar region

cortical spicules

osculum

apopyles

inhalent canals

ostia

radial canals

paragaster

**External View with
Part of Wall Removed**

basal region

Longitudinal Section

inhalent canal   epidermis   mesenchyme

apopyle

choanocytes   prosopyle   radial canal   ostium

Detail of Wall

ostium

parts of
radial canal   inhalent canal

cortical spicules   apopyle

paragaster

Cross-Section

6 mm

© Kendall/Hunt Publishing Company

FIGURE 10.2   Scypha (Grantia) Sponge

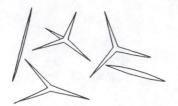

A. Calcareous Spicules
(Class Calcarea)

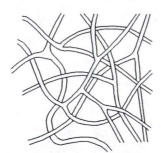

B. Spongin Fibers

micropyle

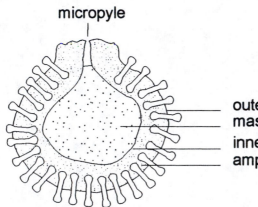

outer membrane
mass of archaeocytes
inner membrane (spongin)
amphidisc (spicule layer)

C. Cross section of a Freshwater
Gemmule

FIGURE 10.3   Sponge Skeletal Structures

Class Hexactinellida (Hyalospongiae)

Example: *Euplectella (Venus flower basket)*
These sponges have funnel shaped bodies formed of 6 rayed siliceous spicules. The latticelike structure of the Venus's flower basket produces an exquisite form. Gently observe the preserved laboratory sample of *Euplectella*.

Class Demospongiae

Example: *Spongia*
The demosponges include the majority of all living sponges. Their skeleton may contain siliceous spicules, spongin protein or combinations of both. The most popular members of this group include the "bath sponges" which are col-

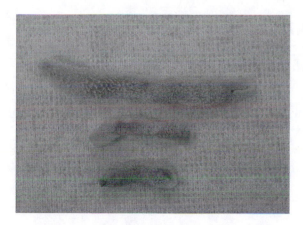

PLATE 10.1   Sycon Body Form

PLATE 10.2   Grantia (Cross-Section)

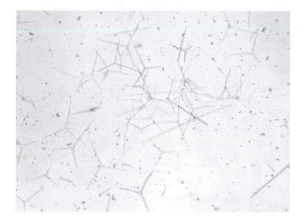

PLATE 10.3   Spicules (Calcareous) (Exterior)

PLATE 10.4   Common Bath Sponges

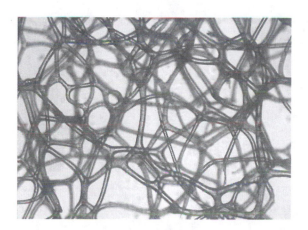

PLATE 10.5   Spongin Skeleton

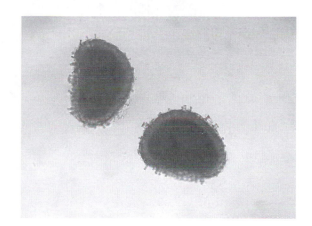

PLATE 10.6   Gemmule

lected by sponge divers, allowed to dry and decompose, leaving their flexible skeletal remains for scrubbing purposes (Plate 10.4). Several bath sponges are available for display. Note the size and texture of such samples as sheep's wool and grass sponges. Obtain a prepared slide or make a wet mount of the skeleton of *Spongilla* and note the protein spongin fibers which intermesh together (Figure 10.3 B, Plate 10.5).

As a special type of survival, freshwater sponges and some marine species produce specialized packets of archeocytes called gemmules. They are surrounded by a tough hard shell embedded with spicules which allow the form to survive through adverse conditions such as drying or freezing. Observe a prepared slide of gemmules and observe the circular form with "spiked" surface (Figure 10.3 C, Plate 10.6).

Questions

1. Why are sponges considered an "unusual" category of animals?

2. Trace the flow of water through the body of a sycon sponge.

3. Describe the basic similarity between sponge choanocytes and members of the protozoan group.

EXERCISE 11   Phylum Cnidaria and Ctenophora

Radiate Animals

Tissue Level of Organization

Learning Objectives:

✔ Describe the function(s) of the various body tissues of the cnidaria.

✔ Explain how digestion in cnidaria differs from that of the protista.

✔ Explain why the division of labor among cells is necessary to increase complexity in body forms.

✔ Explain how polymorphism relates to reproduction in the cnidaria.

✔ Recognize specimens of representative members of the different classes of cnidaria.

Introduction

Phylum Cnidaria

The cnidaria are the lowest **metazoans** (multicellular animals) with true body tissues but lack organs or systems. They include the jellyfish, anemones and corals and are identified by their **radial symmetry** and stinging cells. The **cnido-cytes**, usually along tentacles, possess a triggered **nematocyst** organelle used to inject poison or attach to prey. Food is taken into the **gastrovascular cavity** through a single opening that serves as mouth and anus. The body wall consists of only two well-defined tissue layers, **epidermis** (ectoderm) and **gastrodermis** (endoderm). Between these two layers may be cellular or non-cellular jelly-like **mesoglea**.

Cnidarians illustrate **polymorphism**, different body forms within a complete life cycle. This phenomenon is illustrated both by distinct developmental forms and by differences between individual members or units of a colony. **Polyps** are generally tubular and attached stages while the **medusa** is free-swimming with tentacles dangling from an umbrella-like bell. In some cnidarians the polyp is the only stage while others may have both polyp and medusa stages, or medusa only. Most cnidarians are dioecious and aggregations of gametes arise in tissue masses, or **gonads**. In polymorphic examples the asexual budding polyps alternate with the sexually active medusa forms. A ciliated larva called the **planula** swims and settles onto surfaces to develop into new individuals or polyps.

A significant advance in the process of digestion is apparent in the cnidarians. The digestive system consists of an enclosed chamber (**gastrovascular cavity**) where **extracellular** digestion can take place. Digestive enzymes are released from **epithelionutritive** cells into the chamber and food particles are reduced to smaller particle size by chemical digestion. As a result, extracellular digestion allows the organism to utilize food particles that are larger than it can phagocytize.

Class Hydrozoa

Example: *Hydra*

A living freshwater hydra is a convenient animal for laboratory study. It is a solitary polyp, without a free-swimming medusa phase, and with a greatly reduced mesoglea layer. Because it does not possess a medusa form, this organism is not a "typical" cnidarian. Obtain a living *Hydra* from the jar. Add enough pond water to the culture dish to allow movement and then place the specimen in the dish. Observe the general body form with a hand lens or dissecting microscope (Figure 11.1, Plate 11.1).

Note the normal movements of the body proper and tentacles. Observe reactions to: (1) light tapping and rotation of the dish, (2) lightly touching with a probe, (3) altering the light, (4) adding a drop of water from a culture of living *Daphnia*, and (5) adding a drop of safrain solution. Notice the discharge of the nematocysts.

Obtain different slide preparations of *Hydra*. Some specimens appear to have branches on their sides. These projections are asexual **buds**. In the sexually active *Hydra*, side bulges of **ovaries** or **spermaries** can be seen. From a cross section of *Hydra*, identify: outer epidermis, inner gastrodermis, the mesoglea between the layers and the gastrovascular cavity (**enteron**) toward the center of the ring (Figure 11.2, Plate 11.2, 11.3, 11.4).

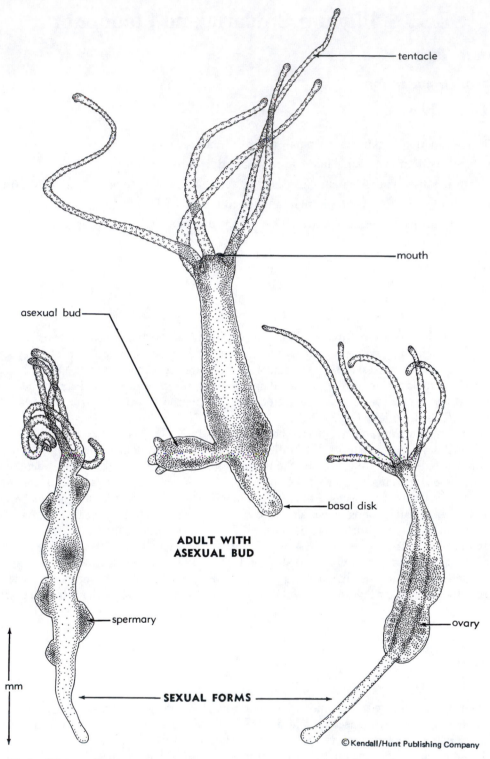

tentacle

mouth

asexual bud

basal disk

**ADULT WITH
ASEXUAL BUD**

spermary

ovary

mm

SEXUAL FORMS

© Kendall/Hunt Publishing Company

FIGURE 11.1   Hydra (External)

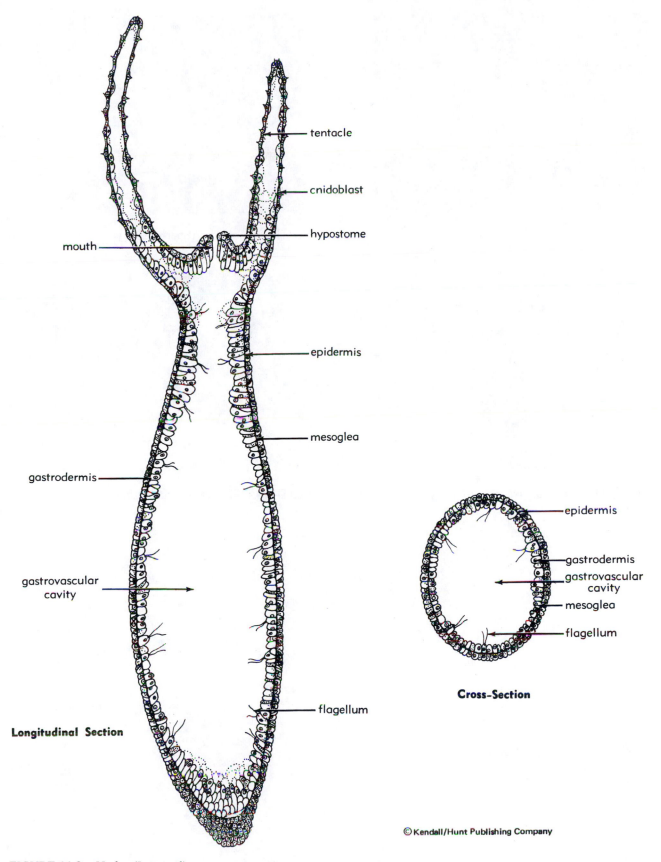

tentacle

cnidoblast

hypostome

mouth

epidermis

mesoglea

gastrodermis

epidermis

gastrodermis

gastrovascular cavity

gastrovascular cavity

mesoglea

flagellum

flagellum

Cross-Section

Longitudinal Section

© Kendall/Hunt Publishing Company

FIGURE 11.2   Hydra (Internal)

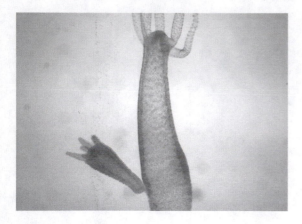

PLATE 11.1   Hydra (With Bud)

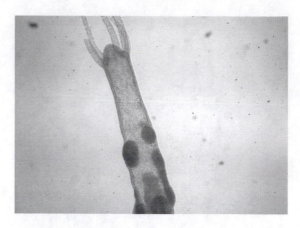

PLATE 11.2   Hydra (Spermary)

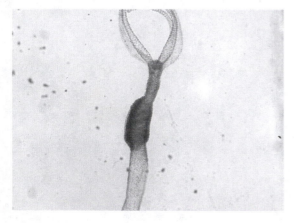

PLATE 11.3   Hydra (Ovary)

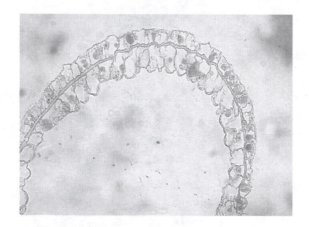

PLATE 11.4   Hydra (Cross Section)

Questions

1. What type of digestion is found in *Hydra?* Compare with a sponge and a protozoan.

2. How does a *Hydra* move from place to place?

3. What special type of cell is unique to cnidarians and what purpose does this type of cell have?

4. Why is the *Hydra* an "unusual" representative of the cnidarians?

PLATE 11.5   Obelia (Polyp)

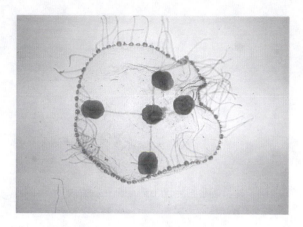

PLATE 11.6   Obelia (Medusa)

Class Hydrozoa

Example: *Obelia*

Polyp

The hydrozoan, *Obelia,* has an asexual stage of development in which it exists as a colony made up of two distinct types of polyps: a food gathering **hydranth** and a reproductive **gonangium**. The tentacled hydranths gather food and share it with the rest of the colony via the gastrovascular cavity. The reproductive gonangium asexually buds medusae from a club-shaped **blastostyle**. The entire colony is covered by a nonliving, chitinous **perisarc**, called the **hydrotheca** when it surrounds the hydranth and **gonotheca** when it surrounds the gonangium (Plate 11.5). Obtain a whole mount slide of the *Obelia* polyp. Observe each of the parts identified in Figure 11.3.

Medusa

Examine a whole mount slide of *Obelia* medusae. Locate the **bell** and the **manubrium** hanging from the **subumbrella** (ventral surface). Hydrozoan medusa also possess a two layered membranous ridge at the subumbrella edge called a **velum**. From the mouth food passes to the enteron and into the **radial canals** and **ring canal**. These structures are lined by gastrodermal cells in which intracellular digestion occurs. Digested food products are passed to amoebocytes that transport them to other body cells. (Figure 11.4, Plate 11.6).

Other Hydrozoans

Located on the demonstration table are preserved specimens of other hydrozoan cnidarians. Members of this class are represented by specimens such as *Gonionemus, Aequorea* and *Polyorchis*. One of the most well known yet strange jellyfish is the *Physalia* or Portuguese Man of War. It consists of a combination of many specialized polyps including reproductive, feeding, and bladders. The numerous, contractile tentacles are capable of severe stings.

Class Scyphozoa

Example: *Aurelia*

Medusa

The class Scyphozoa is represented by jellyfish in which the medusa stage is the most conspicuous stage while the polyp is minute or lacking. The amount of mesoglea in the scyphozoans is greatly increased and provides buoyancy. Specialized muscle cells provide forceful contractions to propel the medusae through the water. The adult *Aurelia* is **dioecious**, being either male or female. Obtain a prepared slide of *Aurelia* medusa (Figure 11.5) and identify the

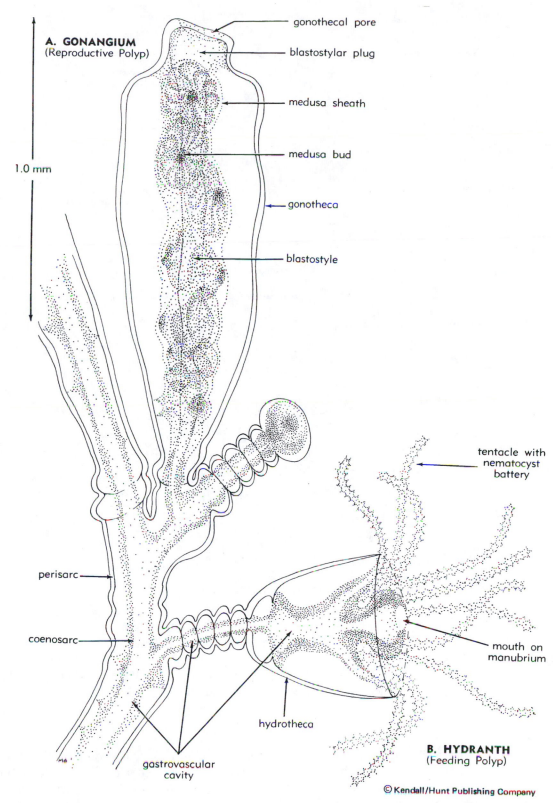

A. GONANGIUM
(Reproductive Polyp)

1.0 mm

gonothecal pore

blastostylar plug

medusa sheath

medusa bud

gonotheca

blastostyle

tentacle with nematocyst battery

perisarc

coenosarc

mouth on manubrium

hydrotheca

gastrovascular cavity

B. HYDRANTH
(Feeding Polyp)

© Kendall/Hunt Publishing Company

FIGURE 11.3   Obelia Polyp Form

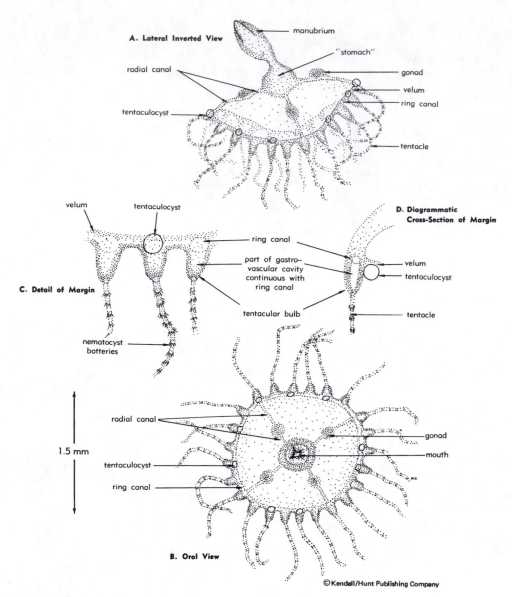

A. Lateral Inverted View

manubrium

"stomach"

radial canal

gonad

velum

ring canal

tentaculocyst

tentacle

velum   tentaculocyst

D. Diagrammatic Cross-Section of Margin

ring canal

part of gastro-vascular cavity continuous with ring canal

velum

tentaculocyst

C. Detail of Margin

tentacular bulb

tentacle

nematocyst batteries

1.5 mm

radial canal

gonad

mouth

tentaculocyst

ring canal

B. Oral View

© Kendall/Hunt Publishing Company

FIGURE 11.4   Obelia Medusa Form

tentacles around the bell edge, mouth, radial canal, and ring canal. Also prominent are: distinctive flaps of the mouth area or **oral arms** used for food gathering, the **gastric pouches** surrounding the mouth area, and the **gonads** closely attached. In scyphozoan medusa, the velum membrane is not present, but the edge of the umbrella is marked by specialized sensory organs for light (**ocelli**) detection and **statocysts** for orientation.

Polyp

Aurelia fertilization produces a ciliated **planula** larva. When it settles, the larva develops into an attached hydra-like polyp called **scyphistoma**. Transverse fission of the top surface in **strobilization** forms a stacked "saucer" appearance of the **strobila**. As each extension or **ephyra** develops, the strobila elongates. When mature, each ephyra breaks off and swims away to develop into a mature medusa (Plate 11.7 A–D).

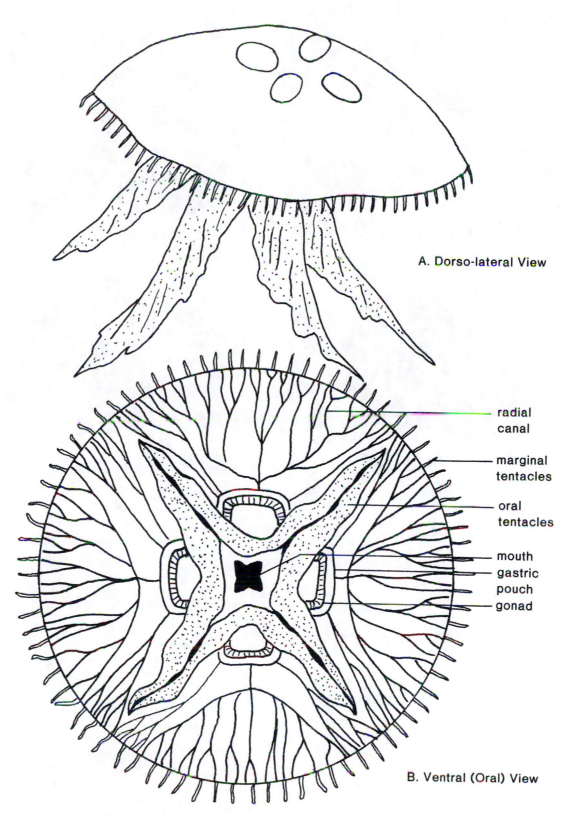

A. Dorso-lateral View

radial canal

marginal tentacles

oral tentacles

mouth

gastric pouch

gonad

B. Ventral (Oral) View

FIGURE 11.5 Aurelia (External)

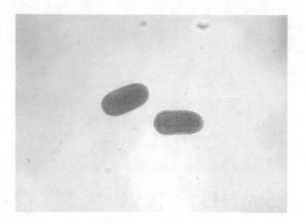

PLATE 11.7A Planula Larva

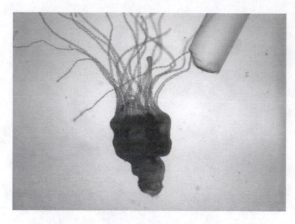

PLATE 11.7B Scyphistoma

PLATE 11.7C Strobila

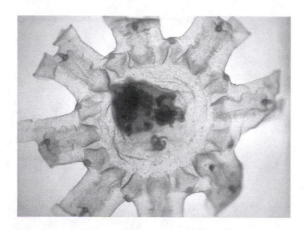

PLATE 11.7D Ephyra

Questions

1. Describe some distinct differences between hydrozoan and scyphozoan cnidarians.

2. What are some of the advantages of having a large mesoglea-filled medusa form?

3. Identify: the site of asexual reproduction in *Aurelia*; the site of sexual reproduction in *Aurelia*.

Class Anthozoa

The class Anthozoa includes the sea anemones and corals. The anthozoans are marine animals that resemble flowers. Therefore, the Greek root word *anthos,* which means flower, is used as a prefix for the class name. They are all sessile polyps but their shapes, colors and forms vary tremendously. Attached at the **pedal disk,** sea anemones have a tubular body with multiple tentacles surrounding an **oral disc.** Observe a sample of *Metridium,* a common sea anemone of both U.S. coasts. Note the thick, muscular body column. If available, observe live anemones in a marine tank to better appreciate their beauty and gracefulness.

Coral members of the anthozoans are also extremely varied. **Hard** or **soft** corals may produce soft swaying forms such as sea fans and sea pens or lay down the calcium "skeleton" of the coral reef. Observe specimens of various corals including stony corals such as brain coral. *Astrangia* forms encrusting colonies along North Atlantic coastlines while staghorn or antler coral forms a recognizable branched colony. *Gorgonia,* or the sea fan, produces colorful shapes in tropical waters. *Renilla* (sea pansy) has a flattened, leaf-like appearance supported by a short attachment stalk. (Figure 11.6, Plate 11.8 A–C)

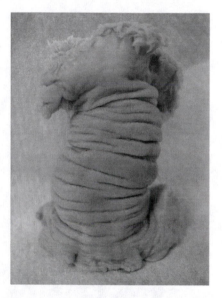

PLATE 11.8 A   Sea Anemone

PLATE 11.8 B   Coral Types

PLATE 11.8 C   Sea Fan

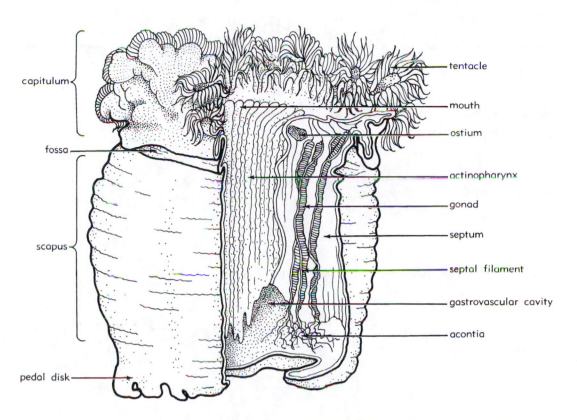

capitulum

fossa

scapus

pedal disk

tentacle

mouth

ostium

actinopharynx

gonad

septum

septal filament

gastrovascular cavity

acontia

LATERAL VIEW (One Quarter Cut Away)

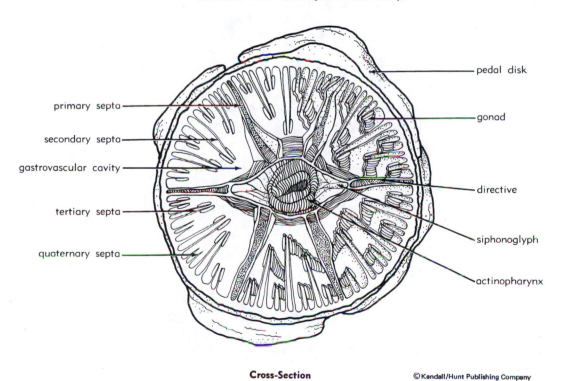

primary septa

secondary septa

gastrovascular cavity

tertiary septa

quaternary septa

pedal disk

gonad

directive

siphonoglyph

actinopharynx

Cross-Section

© Kendall/Hunt Publishing Company

FIGURE 11.6   Sea Anemone

Questions

1. Do anthozoans exhibit polymorphism? Explain.

2. Describe the ecological impact of coral reefs.

3. Do the anthozoans have any economic importance to man? Explain.

Phylum Ctenophora

The Comb Jellies

Introduction

The ctenophores are close relatives of the cnidarians. They possess the characteristic primary radial symmetry in adult form. However, they do not form colonies, do not exhibit polymorphism and lack the stinging nematocysts. They are recognized by comb-shaped plates of cilia, or **ctenes,** used for locomotion. They are monoecious and members may or may not possess tentacles. Many ctenophores, such as *Pleurobrachia* and *Mnemiopsis,* are capable of bioluminescence creating spectacular light shows on dark nights. If possible, observe prepared or jarred specimen of various ctenophores (Figure 11.7).

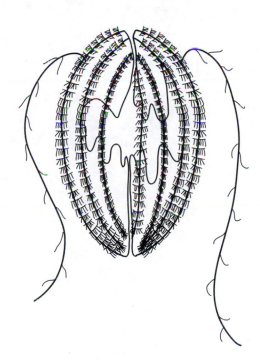

FIGURE 11.7   Ctenophore

Questions

1. Compare and contrast the cnidaria and ctenophora.

2. How is bioluminescence possible?

3. If stinging cells are absent in ctenophores, how do the tentacled forms gather food?

EXERCISE 12   Phylum Platyhelminthes

Flatworms

Organ Level of Development

Learning Objectives

✔ Describe the acoelomate body plan and how this restricts the size of the flatworms.

✔ Explain how the organ level of development in flatworms differs from the tissue level of development in the cnidaria.

✔ Identify representative specimens of the different classes of flatworms.

✔ Describe the life cycle of important human parasites within this group.

Introduction

As the name implies, the platyhelminthes are dorso-ventrally flattened and are called flatworms. They are the first animals to demonstrate the organ level of development, bilateral symmetry, and the simplest form of body organization built largely of mesoderm. They are **triploblastic** or have three germ layers (endoderm, ectoderm, and mesoderm) from which other tissues are derived. The addition of mesoderm allows for the development of a greater variety of tissues which, in turn, makes organ development possible. In the cnidarians, the space between the epidermis and gastrodermis was a noncellular mesoglea. This limited the number of tissues and the variety of body forms. In the flatworms, the mesoderm is a cellular material, including muscle and parenchyma. The development of specialized organs allows for division of labor and a more complicated body form.

Flatworms do not have a **coelom** (body cavity). Body cavities are those spaces lying between the inner surface of the body wall and the outer surface of the digestive tract. Lacking a body cavity, the flatworms are said to be **acoelomate**. The cells that pack the body space between the digestive tract and the body wall do not allow the free flow of fluids, gases, nutrients, and waste to circulate as well as a fluid-filled cavity does. To compensate, the body is flat to allow for adequate diffusion of these materials.

The body of flatworms exhibits bilateral symmetry which allows only one plane of division creating two equal halves. Other phyla, like the cnidarians have more than one plane of division and have radial symmetry. Bilateral symmetry seems to be better adapted for an active life. Because one end of the animal tends to be the forward moving end all the time, locating sensory structures at this end allows the animal to detect stimuli rapidly. This location also promotes brain formation and, in later phyla, head formation. Muscular movement, controlled by a more centralized nervous system, becomes more efficient.

The flatworms include free-living and parasitic forms. When a digestive tract is present, there is a single opening that leads into a "blind" pouch. Most are monoecious and have well developed reproductive systems. They range from a few millimeters in length to extremely large specimens in the parasites. Complicated life cycles have been identified in the parasitic flukes and tapeworms.

Class Turbellaria

Example: *Dugesia (Euplanaria)*

Turbellarians are mostly free-living flatworms and inhabit slow-moving streams and ponds. *Dugesia* is a common laboratory sample and exhibits the typical characteristics of the flatworms. Observe living *Dugesia* with a hand lens or dissecting microscope. Although mostly bottom dwellers, note the muscular contractions which allow a swimming motion. Cilia along the bottom (ventral) surface makes the *Dugesia* move gracefully over surfaces. Look for the distinctive **eyespots** used for light detection and the extended **auricles** along the side. Chemically, these detect the presence of food and tactile receptors interpret information about surface texture. Feed the living *Dugesia* and watch the extension of the muscular **pharynx** as it draws food into the digestive tract (Figure 12.1, Plate 12.1).

Obtain a prepared slide of *Dugesia*. A whole mount slide usually has plain and stained flatworms. Observe the body structures such as eyespots, auricles and pharynx (Figures 12.2, 12.3, Plates 12.2, 12.3). Focus on the darkened digestive

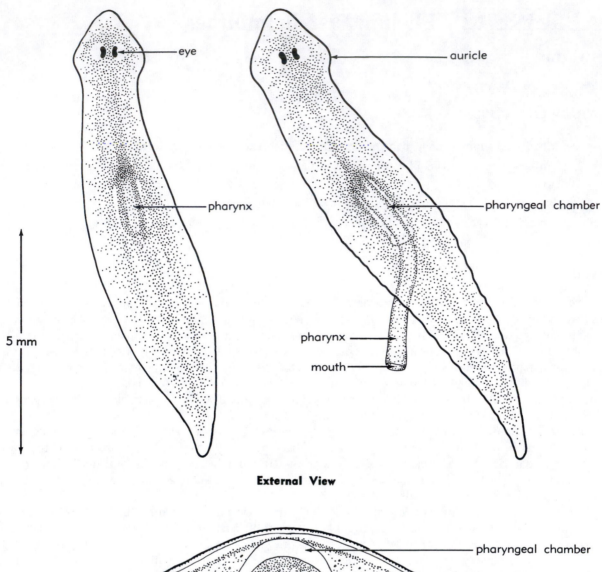

5 mm

External View

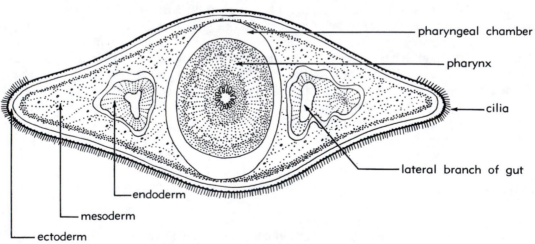

Cross-Section Through Pharynx

FIGURE 12.1   Dugesia (Planaria)

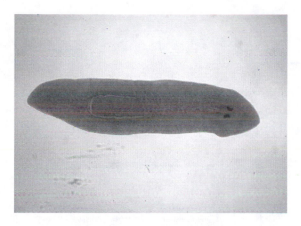

PLATE 12.1   Dugesia (Whole Mount)

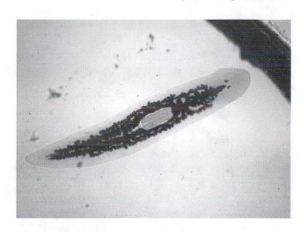

PLATE 12.2   Dugesia (Digestive)

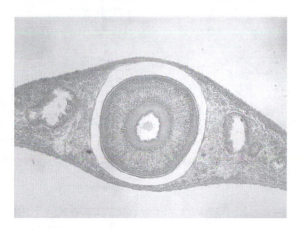

PLATE 12.3   Dugesia (Cross Section)

tract. The pharynx leads into multiple branches of the complex **intestine**. A prepared slide of the cross sectioned *Dugesia* allows observance of internal tissue. Starting at the top surface, identify the ectoderm (epidermis), mesoderm (parenchyma) and ciliated lower epidermis. Three branches of the digestive tract are visible: the larger middle ring is the single, muscular pharynx surrounded by its sheathing chamber and the two smaller rings are the intestinal branches. The endoderm (gastrodermis) lines the digestive branches.

Although not normally visible, the planarians posses a very primitive brain, the **anterior ganglion** and a "ladder" branching of longitudinal nerves. The production of waste material by the multiple layered cells necessitates an excretory process. Many sac-like structures bear tufted cilia (**flame cells**). These produce a cleansing flow of fluid past cells, into tubules exiting through minute pores to the outside surface.

Class Trematoda

Example: *Clonorchis (Opisthorchis)*

All trematodes are parasites and most are internal parasites. *Clonorchis* occurs in human populations of the Far East and Southeast Asia and inhabits the bile duct. Obtain a prepared side of *Clonorchis* and compare its size to that of the planaria (Figure 12.4, Plate 12.4). Note the **anterior sucker** which surrounds the mouth. It serves as a hold-fast organ to attach the fluke to its host. Posterior to the anterior sucker is the **ventral sucker (acetabulum)** located in the middle portion of the body. Like the anterior sucker, the posterior sucker is a hold-fast structure. The **mouth** is surrounded by the anterior sucker. Posterior to the mouth is a muscular **pharynx** which pumps food into the digestive tract. A short **esophagus** extends posteriorly from the pharynx and divides into Y-shaped branches called the **intestinal crura**. Like planaria, *Clonorchis* has no anus: therefore, it has an incomplete digestive tract.

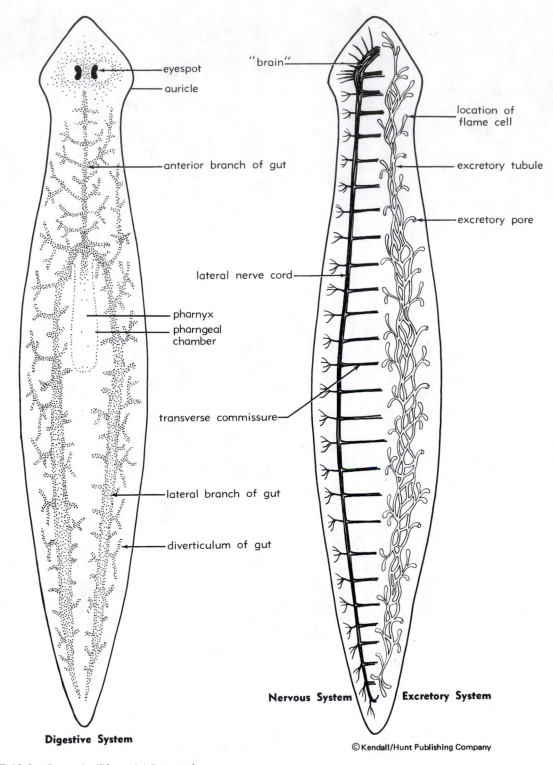

eyespot

auricle

"brain"

location of
flame cell

anterior branch of gut

excretory tubule

excretory pore

lateral nerve cord

pharnyx

pharngeal
chamber

transverse commissure

lateral branch of gut

diverticulum of gut

Nervous System   **Excretory System**

Digestive System

© Kendall/Hunt Publishing Company

FIGURE 12.2   Dugesia (Planaria) Internal

At the posterior end is the **excretory pore** near the excretory **bladder** which it drains. Flukes have both male and female reproductive structures; therefore, they are monoecious but usually cross-fertilize. The female system consists of a single lobed **ovary** about one third of the distance from the posterior end. A large **seminal receptacle** for sperm storage is just posterior to the ovary. Two **yolk (vitelline) ducts** connect the lateral **yolk glands (vitellaria)**. The **oviduct** then extends anteriorly and enlarges into a highly coiled **uterus** filled with encysted eggs. Eggs leave the uterus through the **genital pore** near the ventral sucker. The male system consist of two large lobed **testes** in the posterior re-

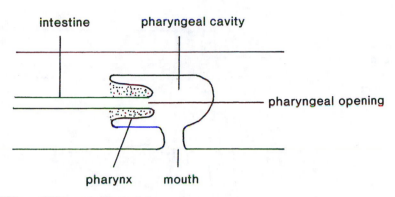

FIGURE 12.3   Lateral View of Pharnyx (Dugesia)

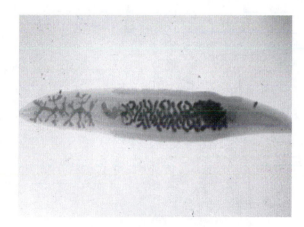

PLATE 12.4   Clonorchis

gion of the fluke. The **vasa efferentia** connects each testis to the single **vas deferens**. The vas deferens extend anteriorly to the male **genital pore** near the ventral sucker.

The life cycle of *Clonorchis sinensis* is shown in Figure 12.5. Fertilized eggs are defecated into the water and must be ingested by a specific snail species. These larval **miracidia** enter the snail's tissue and transform into a bagged **sporocyst** filled with **rediae** (sing; redia). As the rediae move into the snail's liver, they transform again into tailed **cercaria**. These tadpole-looking forms emerge into the water and burrow into the muscle of specific fish species as encysted **metacercaria**. Ingestion of raw or poorly cooked fish with metacercariae releases them into the human digestive tract. Heavy infestations of liver flukes generally results in cirrhosis of the liver and even death.

Other Flukes

If prepared slides or jarred specimens are available, observe other flukes. *Fasciola hepatica,* a very large liver fluke, is found in deer, goats and sheep. It's anatomy is very similar to *Clonorchis,* except for its impressive size. Observe as many body structures as possible using a dissecting microscope. Blood flukes or **schistosomes** invade the veins of human intestines and urinary bladders. A significant difference of *Schistosoma* is that they are dioecious. Larger, broader males have a groove along their length in which the slender female is embraced.

Class Cestoda

Example: *Taenia*

Tapeworms are all internal parasites and usually need two hosts to complete their life cycle. The adult parasitizes the digestive tract, but the immature forms may be found in various organs. Since the adult form lacks a digestive tract, the parasite simply absorbs the products of the host's digestive process. By locating in the digestive tract, the eggs made

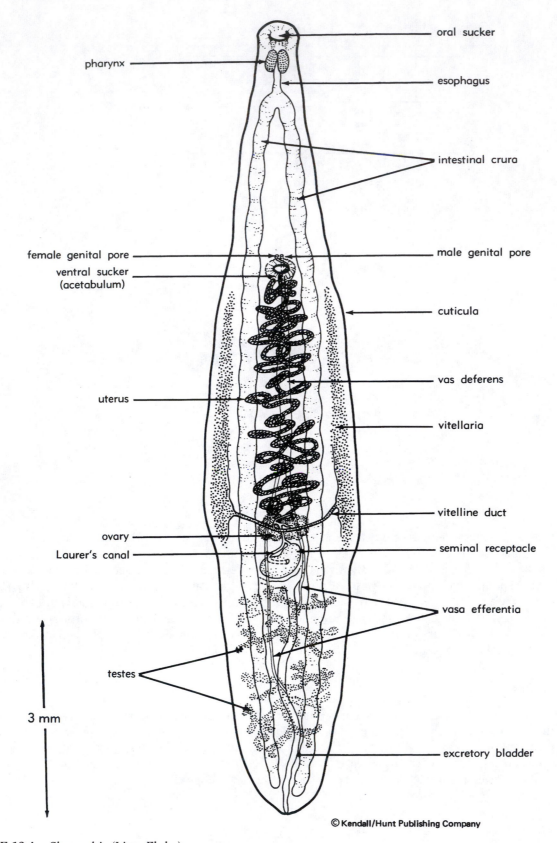

oral sucker

pharynx

esophagus

intestinal crura

female genital pore

male genital pore

ventral sucker
(acetabulum)

cuticula

vas deferens

uterus

vitellaria

vitelline duct

ovary

seminal receptacle

Laurer's canal

vasa efferentia

testes

3 mm

excretory bladder

© Kendall/Hunt Publishing Company

FIGURE 12.4   Clonorchis (Liver Fluke)

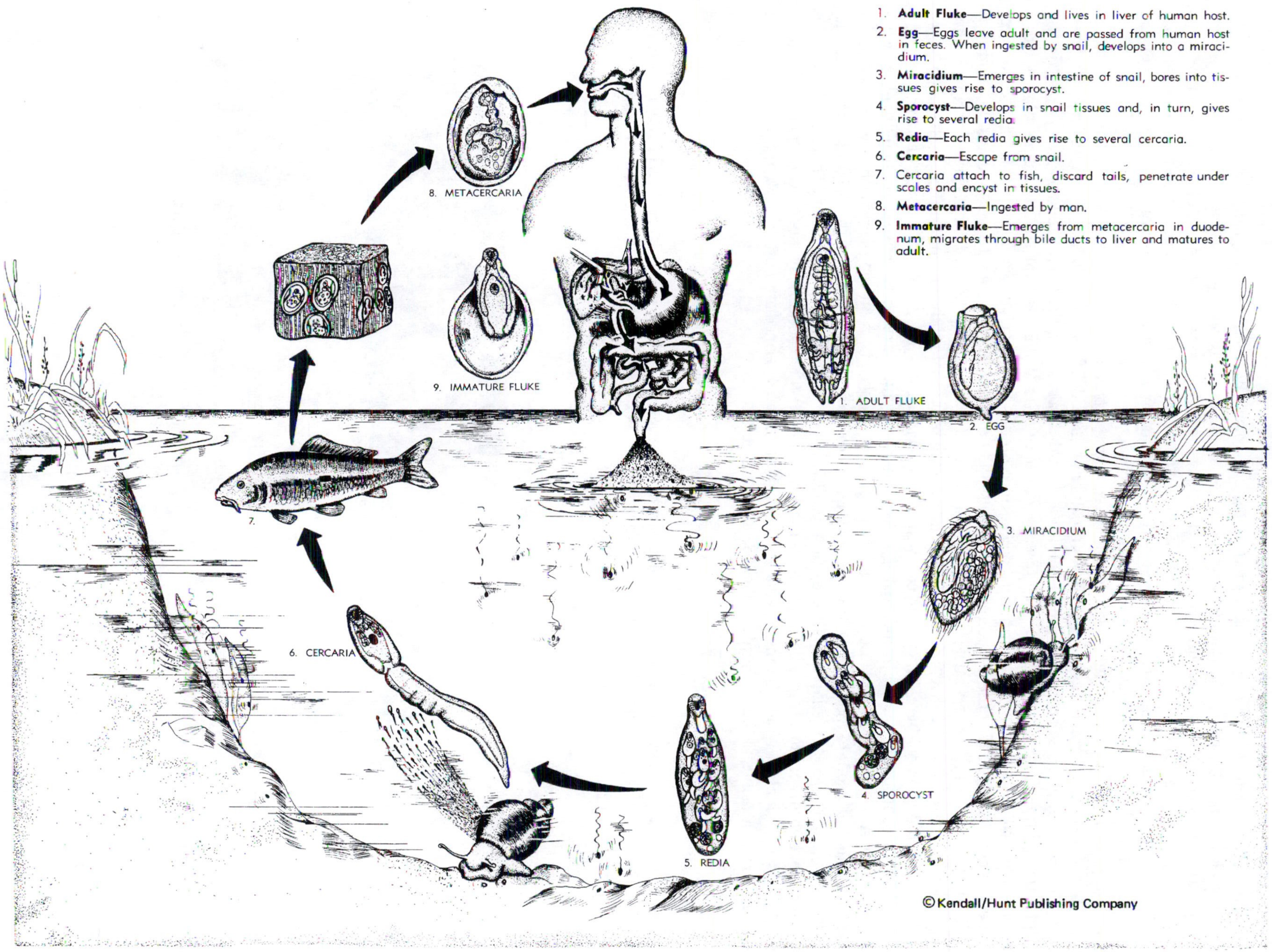
1. Adult Fluke—Develops and lives in liver of human host.
2. Egg—Eggs leave adult and are passed from human host in feces. When ingested by snail, develops into a miracidium.
3. Miracidium—Emerges in intestine of snail, bores into tissues gives rise to sporocyst.
4. Sporocyst—Develops in snail tissues and, in turn, gives rise to several redia.
5. Redia—Each redia gives rise to several cercaria.
6. Cercaria—Escape from snail.
7. Cercaria attach to fish, discard tails, penetrate under scales and encyst in tissues.
8. Metacercaria—Ingested by man.
9. Immature Fluke—Emerges from metacercaria in duodenum, migrates through bile ducts to liver and matures to adult.
1. ADULT FLUKE
2. EGG
3. MIRACIDIUM
4. SPOROCYST
5. REDIA
6. CERCARIA
7.
8. METACERCARIA
9. IMMATURE FLUKE
©Kendall/Hunt Publishing Company

FIGURE 12.5 Clonorchis Life Cycle

by the adult tapeworm have an exit to the environment where additional hosts may be infected. By locating in organs that may be eaten, the immature form gains access to a new host.

On prepared slides of *Taenia,* various sections of the tapeworm have been mounted. Progressing from the smallest to largest sections, identify the various parts. Since the adult tapeworm lacks a digestive system, most of the internal structures belong to the reproductive system. The body consist of three regions: the anterior **scolex**, a short **neck** region and a long ribbon-like body known as the **strobila** (Figure 12.6). The scolex is a permanent structure and has **hooks (rostellum) and/or suckers** at the anterior end. These structures serve as holdfast organs. At the posterior end of the scolex is the **neck** which continuously buds new sections or **proglottids**. Like the fluke, the tapeworm is monoecious so each proglottid contains male and female reproductive organs.

The most anterior proglottids are known as **immature** proglottids whose reproductive structures are just developing. Consequently, these structures are ill-defined and hard to see. Along the lateral edge of each proglottid is a lightly stained line, an **excretory canal**. The excretory canals connect at the posterior end of each proglottid with a **transverse canal**. These empty waste from the body at the posterior end of the tapeworm. One or two large swellings or **genital pores** are located at the midpoint of each proglottid. From the genital pore extends two ducts. The more anterior duct is the **sperm duct** of vas deferens and the posterior duct is the **vagina**. The vas deferens branches into smaller ducts or **vasa efferentia** that collect sperm from the many small testes that fill the proglottid. The vagina extends posteriorly toward the **ovary** where it then connects with a short oviduct, the yolk glands, and a long slender **uterus**.

Gravid proglottids are filled with encysted eggs. The other reproductive organs are usually difficult to observe because of the numerous eggs. These fertilized eggs develop into six-hooked larva known as **onchospheres**. Passed out in the feces and ingested into the digestive tract of the **intermediate host**, these larva attach to the wall of the intestine. Obtain a prepared slide of the larva **cysticerus** (Figure 12.7, Plate 12.5 A&B) or **bladderworm**. This stage in the beef and pork tapeworms produces "measly" meat. Note the distinct head region but usually the neck and body are reduced in size and invaginated. When the intermediate host or parts of the intermediate host are eaten, the bladderworm evaginates and attaches to the intestinal wall of the **definitive** host.

Other Tapeworms

Located on the demonstration table are examples of other tapeworms. Observe jarred specimens of *Taenia saginata* (beef) or *Taenia solium* (pork) and note the extensive length of an extracted adult. If slides are available, note the similar structures in *Dipylidium caninum,* a tapeworm commonly infesting dogs.

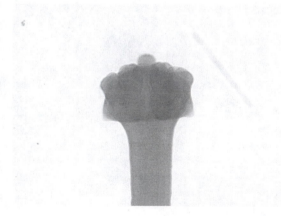

PLATE 12.5A   Scolex

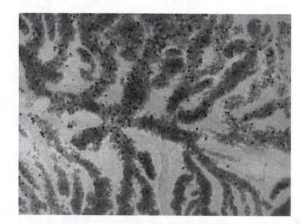

PLATE 12.5B   Gravid Proglottid

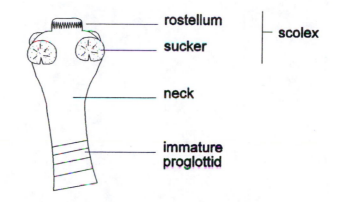

A. Scolex and Neck Region

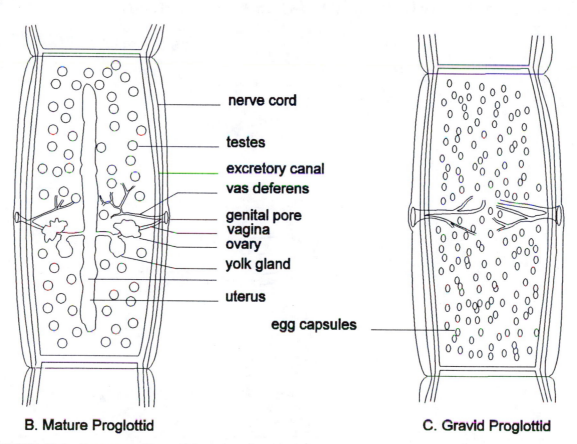

B. Mature Proglottid

C. Gravid Proglottid

FIGURE 12.6   Tapeworm (General Structures)

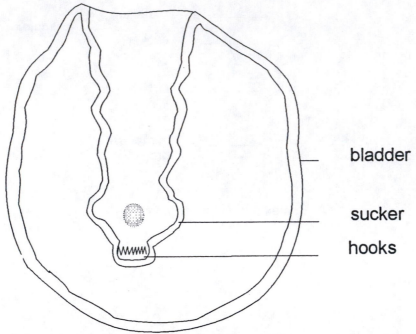

bladder

sucker

hooks

A. Invaginated Cysticercus (Bladder Worm), Longitudinal Section.

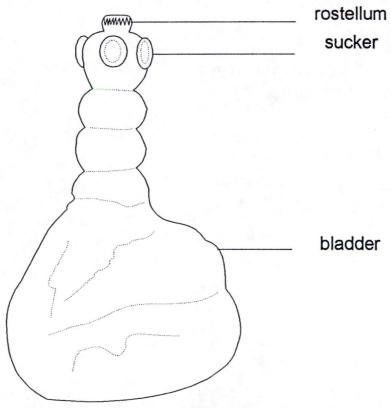

rostellum

sucker

bladder

B. Evaginated Cysticercus (Bladder Worm).

FIGURE 12.7   Cysticercus (bladderworm).

Questions

1. What structures differentiate the anterior end of planaria? Of liver flukes? Of tapeworms?

2. What type of symmetry do the flatworms possess?

3. How is locomotion accomplished in planarian worms?

4. How is food ingested in the planarians?

5. How do the flukes and tapeworms get food without digestive organs?

6. What is a flame cell? Explain its function.

7. How could infections by flukes and tapeworms be halted?

8. What is the advantage of being a monoecious parasite?

EXERCISE 13 Phylum Nematoda

Roundworms

Complete Digestive Tract

Learning Objectives

✔ Describe how the organ-system level of complexity in this group differs from that of the flatworms.

✔ Explain the importance of the pseudocoelom and its advantages over the acoelomate body plan.

✔ Identify representative specimens of the different classes of nematodes.

✔ Recognize the life cycles of parasitic nematodes and identify preventive techniques of noninfestation.

Introduction

The phylum Nematoda is the first group of animals to be studied that has a **complete** digestive tract, a tube within a tube body plan. This allows for food to be broken down and absorbed as it moves along the tract from the anterior to the posterior end of the animal. A muscular **pharynx** at the anterior end ingests the food and the undigested portion passes out a second opening known as the **anus**. Therefore, food does not have to be regurgitated to be eliminated. Nematodes also have a type of body cavity known as a **pseudocoelom**. This fluid-filled chamber allows nutrients absorbed from the gut tract to freely circulate from front to rear within the worms. No cell membranes act as barriers to slow the movement of nutrients as in the flatworms. Consequently, the pseudocoelom serves as a simple type of circulatory system. It also functions as a hydrostatic skeleton to stiffen the body. Being **bilaterally symmetrical**, the body is stream-lined for movement. These two factors allow for efficient use of the muscular system.

Nematodes are chiefly free-living and microscopic. They range is size from microscopic to a meter in length. They occupy a great number and variety of habitats in soil, mud, and fresh or salt water. Some soil nematodes, however, are plant parasites and cause millions of dollars in agricultural damage annually. Many animals, including humans, are parasitized by nematodes such as pinworms, *Trichinella,* numerous filarial worms, and several species of hookworms.

Example: *Ascaris lumbricoides*

Ascaris is not a typical nematoda. It is used because of its large size and availability. Most of its pseudocoel is filled with a tremendously elongated coiled reproductive system. The female is capable of producing 200,000 eggs daily. Although infection by Ascaris is **direct,** the path of the larva through the body is complicated. Ingested eggs are swallowed, hatch in the intestine but then burrow through the wall and enter the bloodstream. At the lungs, the immature worms break out into the alveoli, are coughed up, swallowed and finally reach the intestine for anchorage.

Obtain a preserved specimen of *Ascaris* and place in a dissecting pan. Most nematodes are dioecious and there are usually external differences sufficient to distinguish them apart. Look for the obvious size differences between males and females. The male is smaller and the tail end is conspicuously curled with a pair of copulatory **spicules.** On the larger female find the **head** with three **lips** surrounding the **mouth.** The largest lip is dorsal, and the other two are ventro-lateral (Figure 13.1). The anal end is somewhat larger than the anterior end.

Internal Anatomy

After selecting a female worm, dissect the worm as follows. Start to slit the animal open along the its length beginning at the head region. The internal organs are very soft so barely insert the point of a dissecting needle just under the body covering and pull the tip along. The body wall will split open easily. Using large dissecting pins, pin the body wall on either side exposing the organs. Identify the muscular **pharynx** and the ribbon-like **intestine** which extends to the anal opening. Look for a "wishbone" like structure about one third down the length. The short **vagina** is attached to the **genital pore** and splits into two large **uterus** branches. The eggs produced in the terminal **ovaries** move through the coiled, tubular **oviducts.** As they pass into the uterus, stored sperm fertilize them.

Examine a prepared slide of male and female *Ascaris* worms. In the cross section of the male, note the flattened **intestine** and the numerous **sperm** ducts or **vas deferens.** Identify the **cuticle** covered body wall, the **muscle** layer beneath and the clear cavity of the **pseudocoelom.** In the cross section of the female, similar structures will be evident. However, the two large **uteri** filled with eggs are usually the most predominant structures. The smaller circular structures are the **oviducts** and **ovaries** (Figure 13.2, Plate 13.1 A, B).

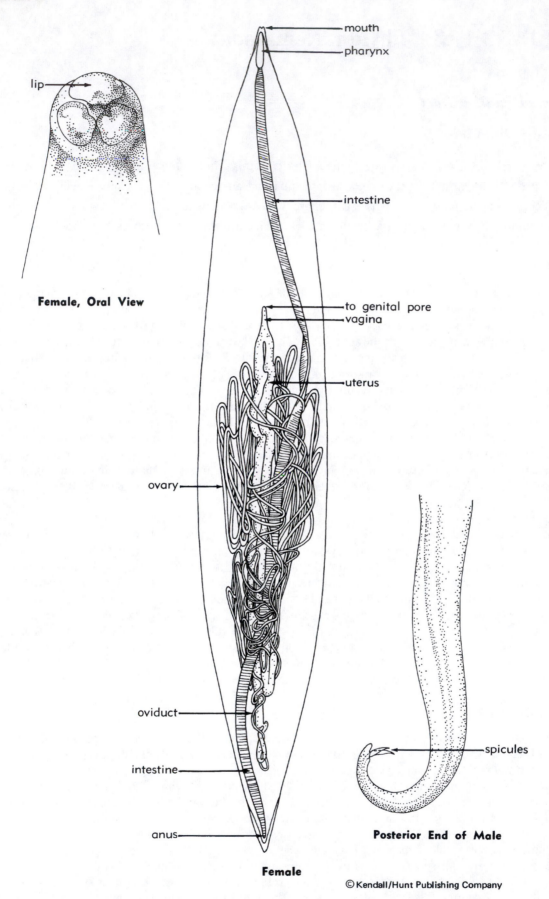

lip

Female, Oral View

mouth

pharynx

intestine

to genital pore

vagina

uterus

ovary

oviduct

intestine

anus

Female

spicules

Posterior End of Male

© Kendall/Hunt Publishing Company

FIGURE 13.1 Ascaris Anatomy

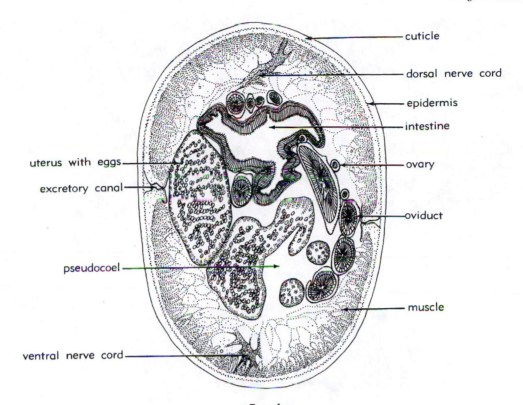

Female

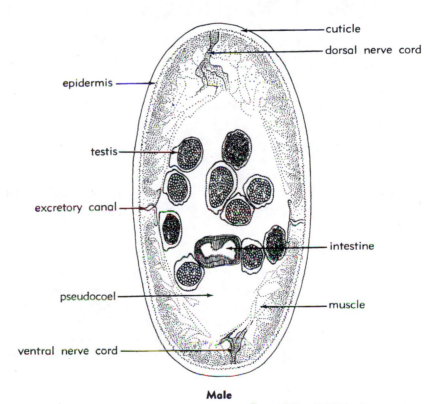

Male

FIGURE 13.2 Ascaris (Female and Male)

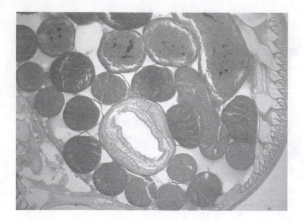

PLATE 13.1A   Ascaris (Male)

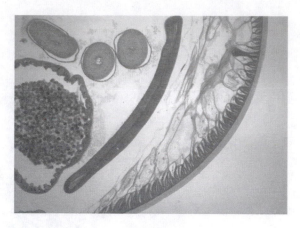

PLATE 13.1B   Ascaris (Female)

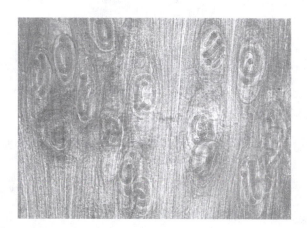

PLATE 13.2   Trichinella

Example: *Trichinella*
Examine a prepared slide of *Trichinella* and observe under low-power magnification. Look for the encysted worms rounded up within the muscle fibers. Humans become infected with *Trichinella* by eating contaminated and poorly cooked pig meat (Figure 13.3, Plate 13.2).

Example: *Enterobius vermicularis*
The pinworm or seatworm is a cosmopolitan parasite of the human intestinal tract with a direct life cycle including self-contamination. Upon ingestion or inhalation of embryonated eggs, development continues in the intestinal tract where attachment, mating and egg production occurs. Females migrate to the anal opening at night to deposit the eggs on the skin surface. This accounts for the easy recontamination by children as they scratch the irritated perianal area. Inhalation of eggs on bedclothes or sheets is also a common infestation mode by adults attending infected children. Observe a prepared slide of *Enterobius* and find the light pointed tail region which gives it one of its common names (Figure 13.3).

Example: *Necator americanus*
These hookworms possess a slightly curved anterior end whose mouth area has cutting plates. Attached to the intestinal wall they feed on a blood meal resulting in symptoms of anemic conditions. Infestation by juveniles in the soil usually occurs through skin surfaces, primarily the soles of the foot. Migration through the body is similar to the *Ascaris* trek. Look at a prepared specimen of *Necator* or *Ancylostoma*. Compare it to *Enterobius* and be able to distinguish between them (Figure 13.3).

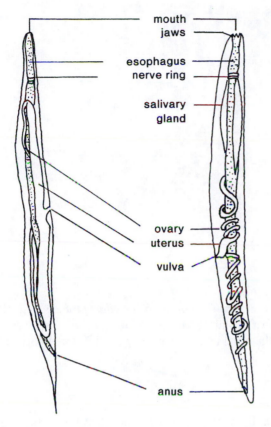

A. Pinworm, Female

B. Hookworm, Female

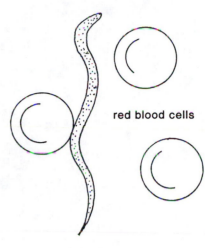

C. Microfilaria of
Dirofilaria immitis

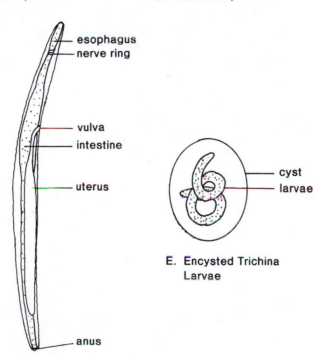

D. Trichina Worm, Larvae

E. Encysted Trichina
Larvae

FIGURE 13.3   Parasitic Roundworms

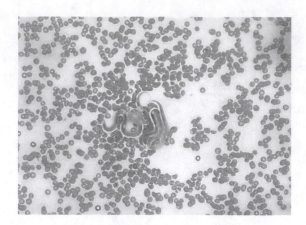

PLATE 13.3   Dirofilaria

Example: *Dirofilaria*

The filarial worms invade fluid tubes such as blood and lymph vessels. The common *Dirofilaria* infects dogs through the bite of mosquitoes. The larval forms are identified in prepared slides of blood tissue and can be seen as slender worm forms between the blood cells. Matured worms live in the heart and lung arteries resulting in enlarged organs and eventually heart and lung failure (Figure 13.3, Plate 13.3).

Other Nematodes

If available, watch living vinegar eels (*Turbatrix*) move about in the vinegar solution. They thrive on the bacteria and yeast there. Another significant nematode is *Caenorhabditis*. It has become a classic research animal in studies of genetics and developmental control. Because of its limited number of body cells, scientists have been able to map the source of all body cells in the adult worm and are studying the genetic control of differentiation.

Questions

1. What type of symmetry is found in the nematodes?

2. What type of diet do the different types of nematodes have?

3. Give a comparison between flatworms and nematodes.

4. Explain the term dioecious. Is this an advancement over monoecious animals?

5. What are the advantages of a pseudocoelom?

6. How might a heavy infestation of *Trichinella* affect muscle function?

7. What is the most practical medical test/exam to determine the presence of intestinal nematodes?

8. Identify safety precautions to prevent infection by specific parasitic nematodes.

EXERCISE 14 Phylum Mollusca

Soft Body Form

Learning Objectives

- ✔ Describe the general body plan of the mollusc.
- ✔ Describe the structure and function of the mollusc shell.
- ✔ Recognize representatives of selected molluscan classes.

Introduction

The phylum Mollusca is a well-defined but extremely diverse group of animals. The body is bilaterally symmetrical, unsegmented, and soft with a **mantle** covering. The mantle's epithelium secretes a calcareous **shell** in most members of the group. Most also have a ventral foot, a dorsal, visceral **body mass**, true **coelom**, and gills. Some like the oysters have no head or foot. Adapting to different means of nutrition and methods of locomotion has created a wide variety of body forms among the molluscs. The squid specializes in speed and predation, the clam for filtering detritus, and the snail for gliding and grazing. They have adapted to marine, aquatic, and terrestrial habitats. Molluscs have well developed excretory, respiratory and reproductive systems. The **open** circulatory system pumps blood through vessels but also allows blood to bathe tissues directly in **sinus** areas. The coelom cavity is reduced to an area just around the heart. Most molluscs are dioecious and they reproduce sexually.

Class Bivalvia

Example: *Unio or Venus*
Bivalves are molluscs such as mussels, clams, oysters, and scallops that have two shells or **valves**. The shells are tightly closed by well-developed **adductor muscles**. Bivalves feed by filtering food from an internally controlled water flow.

Shell Features

Look at the cleaned shell of a bivalve clam. Determine the body orientation by looking for the swollen area known as the umbo. The umbo is the oldest part of the valve and is on the **anterior**, **dorsal** side of the clam. The extended end away from the umbo is the **posterior** end. The open edge of the clam opposite the hinge is the **ventral** margin. Note the **lines of growth** on the valves. The wide band of each represents a season's growth (Figure 14.1, Plate 14.1A-B).

By placing the proper shell "palmed" in the hand, place the thumb on the umbo with the short, anterior end pointed upwards. This should place the ventral margin at the fingertips. The correctly palmed shell can then be designated **right** or **left** valve. (If the shell does not "set" properly, switch it to the other hand and try again.) Look at the various layers of

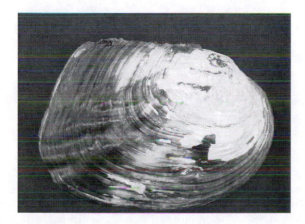

PLATE 14.1 A   Bivalve Shell (Exterior)

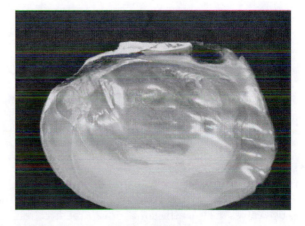

PLATE 14.1 B   Bivalve Shell (Interior)

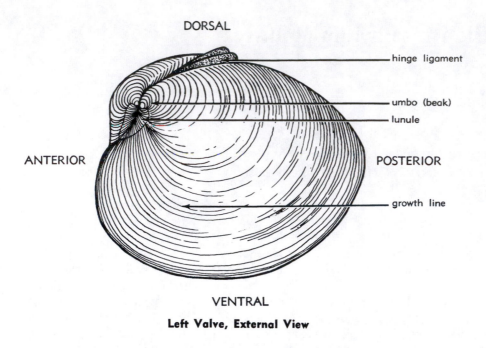

Left Valve, External View

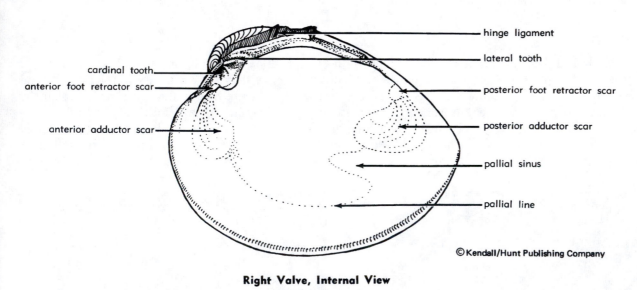

© Kendall/Hunt Publishing Company

Right Valve, Internal View

FIGURE 14.1   Bivalve Shell

the shell surface. The outer is the thin, horny layer or **periostracum**. The next layer is the **prismatic** layer and may be seen when the periostracum is scraped or chipped away. The innermost layer is the **mother-of-pearl** or **nacre**.

On the internal shell surface, just at the hinge are the longitudinal **lateral teeth**. These teeth prevent movement of the two valves in either the dorsal or ventral direction. Just inside and below the umbo are the pointed **cardinal teeth**. These teeth prevent any anterior or posterior movement in the valves (Figure 14.1). Powerful **adductor muscle scars**

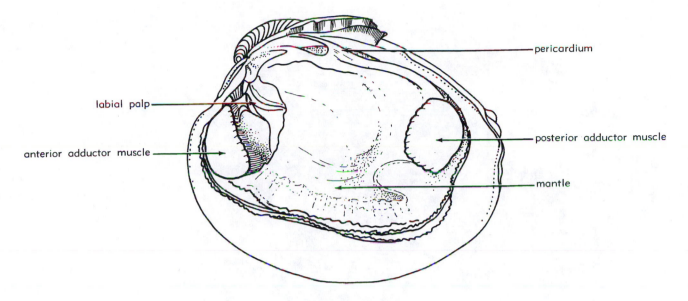

Left Valve Removed, Mantle Intact

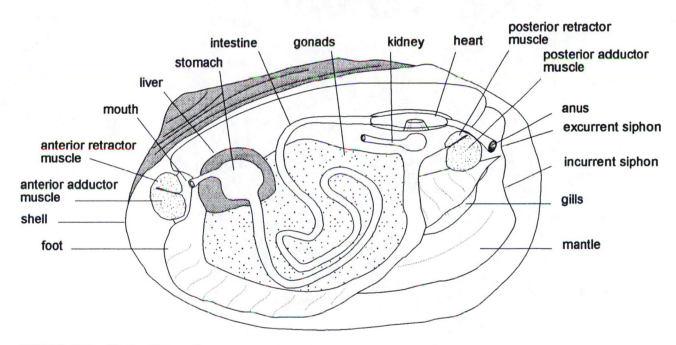

FIGURE 14.2   Bivalve (Internal)

can be seen on either side of the hinge and the pallial line follows the curve of the ventral shell edge. This line is the scarred surface on the valve where the mantle edge attaches.

Obtain a fresh-water clam and place it in the dissecting pan. The shell is held apart by a wooden peg. Remove the peg and insert a heavy bodied knife on either side of the umbo, severing the muscles. Pry open the shell and scrape the soft body parts into the bottom half. Examine the ragged edges of the torn **adductor** muscles and note their thickness. Focus on the soft body and fold back the soft, thin membrane or mantle. Two openings at the **posterior** end of the clam include the **incurrent siphon** and the **excurrent siphon**. Water enters through the incurrent siphon, circulates around the mantle cavity, then through the gills. Filtered water is discharged through the excurrent siphon. Food particles and oxygen are removed as water circulates through the clam (Figure 14.2 A-B, Plate 14.2).

Remove the exposed mantle half, carefully cutting it with scissors. Below it lies the **visceral mass**, and foot. A pair of striped **gills** lie on each side of the visceral mass. Locate the slit-like **mouth** opening between the foot and the

A. Clam, intact organs.

B. Clam, dissected organs.

PLATE 14.2   Clam Dissection

anterior adductor muscle. It is bordered by a pair of small triangular **labial** (oral) **palps** that aid in passing food-laden mucus strings into the slit-like *mouth*.

Locate a cavity directly in front of the posterior adductor muscle and just below the hinge. The wall of this cavity is a thin and semitransparent membrane. Cut through the membrane to expose the **pericardial cavity** in clams, which contains the **heart**. It is a triangular structure wrapped around a portion of the digestive tract called the **rectum**. Below and lateral to the pericardial cavity is a pair of dark organs, **kidneys**, which drain the coelom.

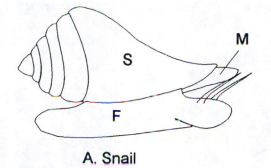

A. Snail

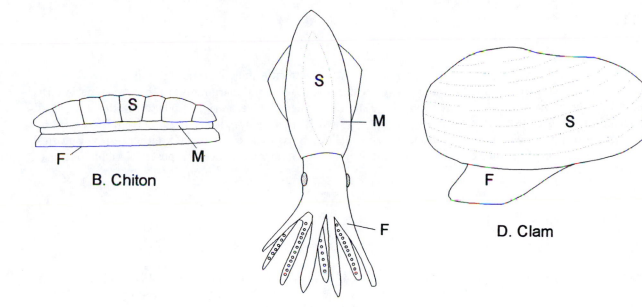

B. Chiton

C. Squid

D. Clam

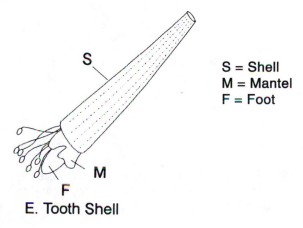

S = Shell
M = Mantel
F = Foot

E. Tooth Shell

FIGURE 14.3 Classes of Molluscs

PLATE 14.3   Chiton

PLATE 14.4   Tooth Shells

Next, remove the body from its remaining valve. Trim away the mantle and gill tissue with scissors. Using a sharp scalpel, dissect the foot through the visceral mass by **cutting along the ventral margin toward the dorsal margin**. Dissect only the visceral mass. Locate the mouth and the short **esophagus**. The esophagus leads to the **stomach**, an enlarged sac. Masses of greenish tissue on either side of the stomach are **digestive glands**. Leading from the stomach is the twisted tube or **intestine** that is embedded in a mass of yellowish tissue called the **gonads** (either testes or ovaries). The rectum is a portion of the intestine that enters the coelom. Trace the rectum over the posterior adductor muscle where it ends at the **anus**.

Class Polyplacophora

Chitons

Chitons are members of the class Polyplacophora (Figure 14.3, Plate 14.3). They have an elliptical body with a shell containing a mid-dorsal row of eight broad plates. Chitons have a large, flat foot surrounded by a row of gills. Pick up and observe the ventral and dorsal surfaces of preserved chitons. Identify the overlapping plates that give the class its name.

Class Scaphopoda

Tooth or Tusk Shells

These elongated, single shelled mollusks inhabit both shallow and deep oceans. With an opening at both ends of the tubular shell they can burrow headfirst into sand or mud while the siphons circulate water through the shelled surface exposed to the water. Look at some tusk shells and note the delicate shell formations (Figure 14.3, Plate 14.4).

Class Monoplacophora

Members of this group were identified by fossil shells and thought to be extinct. However, living specimens of *Neopilina* are now recognized as living examples of this class. The shell is a circular, cap-like dome that protects the soft body. If *Neopilina* specimens are available, note their similarities to limpet shells.

Class Gastropoda

Snails and slugs are universal molluscs that are members of the class Gastropoda. The shell is usually coiled, reduced, or absent. There is a distinct head, commonly with eyes and tentacles. Most gastropods also have a rasping structure or **radula** within the mouth area. Look at a prepared slide of a snail radula and note the overlapping ridges which allow the

PLATE 14.5   Snail Radula

PLATE 14.6   Gastropod Shells

snail to scape algae or other debris from rock surfaces (Plate 14.5). Locomotion is by a large, flat muscular foot. Many of the marine gastropod snails are known to produce harmful toxins that are injected into prey. Nudibranchs and sea slugs are naked gastropods, often with extended projections on their dorsal surface for gas exchange. Some brightly colored species warn of the dangerous poisons contained in those projections. Observe various specimens of this class including: conch shells, slugs, land snails (*Helix*), nudibranchs, cone shells, abalone and limpets (Figure 14.3, Plate 14.6).

Class Cephalopoda

The class Cephalopoda is represented by the squid and octopus. The squid has a large head which is pointed with fins and conspicuous eyes. The mouth is surrounded by ten tentacles. The octopus has a rounded head with large eyes, but no fins. The mouth is surrounded by eight arms. The chambered *Nautilus*, an exquisite deep-diving cephalopod has a coiled shell separated into compartments. When new sections are added, the soft body moves forward into the completed chamber and the remaining fluid-filled chambers are pressurized for buoyancy. Look at the various specimens of the cephalopod group on display including: the cuttlefish (*Sepia*), octopus, squid and *Nautilus* (Figure 14.4, Plate 14.7 A-B).

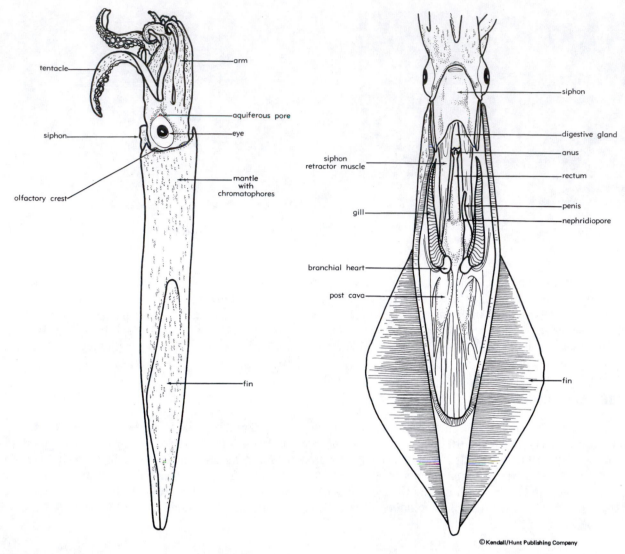

FIGURE 14.4   Squid Anatomy

© Kendall/Hunt Publishing Company

PLATE 14.7 A   Squid

PLATE 14.7 B   Nautilus

Questions

1. Why are the clam adductor muscles larger than the retractors?

2. Describe the various modes of ingestion found in molluscs.

3. What are the different layers of the molluscan shell and their function?

4. Of what economic importance are the molluscan members?

5. How is the octopus eye like a human eye?

EXERCISE 15 Phylum Annelida

Segmented Worms

True Coelomates

Learning Objectives:

✔ Explain how metamerism in the annelids has contributed to the complexity of the annelid body form.

✔ Explain how regionalization of the internal organs has contributed to the complexity of the annelid body plan.

✔ Identify representative specimens of the various annelid classes.

Introduction

The phylum Annelida is a group of segmented worms that include earthworms, marine bristle worms, and leeches. The annelids have a true coelom (**eucoelom**) and a **closed** circulatory system. It is the simplest of the eucoelomate animals, yet, it possesses important organ systems characteristic of all higher forms.

Perhaps the most outstanding feature that separates annelids from other worms is their **segmentation**. This repetition of body parts is also called **metamerism** and is internal as well as external. Metamerism is important because it has provided a redundant supply of organs, each of which is specialized for new and different functions. In short, metamerism provided a means by which a greater variety of organizational plans was possible. In spite of the possibilities for advanced forms, only in arthropods and chordates do we see the extent of body forms that metamerism makes possible. The compartmentalization of the body cavity by septa makes the hydrostatic skeleton more efficient. Aiding this is the presence of two muscle groups (longitudinal and circular) in the body wall and the fusion of the nervous system into one unit.

Class Oligochaeta

Example: *Lumbricus terrestris*
Oligochaete earthworms live in moist, humus soil. They are nocturnal animals and sometimes spoken of as "night crawlers." Other oligochaete characteristics are internal and external segmentation, a reduced head or head lacking, paired setae per somite, monoecious, and a **clitellum**.

External Anatomy

Obtain a preserved earthworm, hand lens, and a dissecting pan. Note the obvious separation of the body into **segments** or **metameres**. The mouth and **prostomium** are located at the anterior end, and the anus at the posterior end of the animal. The **clitellum** is a girdle-like structure about one-fourth of the worm's length from the anterior end. It is important in the formation of mucus during copulation and the secretion of the **cocoon** or **egg case**. The dorsal surface is darker than the ventral surface due to a faint dark **dorsal blood vessel** that runs down its length. With the earthworm between your fingers, pull gently. Chitinous bristles, or **setae** should be felt along the ventral and lateral surfaces. Each segment has four pairs of these setae. Use a hand lens or dissecting scope to see them in a dried worm specimen. As the skin contracts, they are more clearly visible along the surface (Figure 15.1, Plate 15.1).

Locate paired openings on the fourteenth and fifteenth somites. The openings on the fourteenth pair are **oviduct** openings and the fifteenth pair are **sperm duct openings** (vas deferens openings).

Internal Anatomy

Position the earthworm dorsal side up in the pan. To stabilize the specimen, place a pin through the middle of the animal about **midway** between the anterior and posterior end. Stretch the worm slightly and pin the worm through the **fourth** somite from the anterior end. With a scalpel, make a small slit in the dorsal surface **posterior** to the clitellum

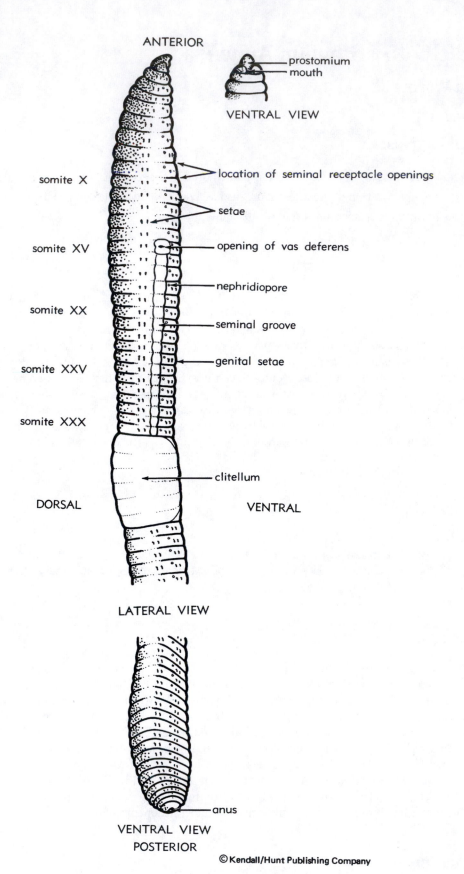

FIGURE 15.1   Earthworm (External)

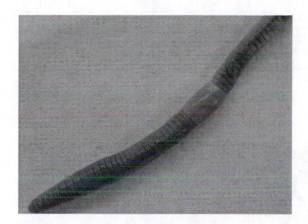

PLATE 15.1   Eathworm (External)

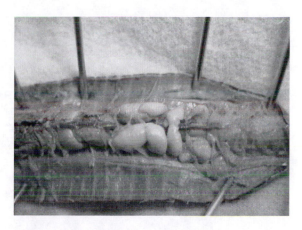

PLATE 15.2   Earthworm (Internal)

and continue the incision along the mid-dorsal line to the fourth somite. Avoid damaging the internal structures by taking care to just cut through the body wall. Beginning at the point of the **initial** incision, open the worm by pinning back the cut edges of the body wall for approximately one inch. Then cut the internal partitions, or septa, along each side of the cut with a **probe**. Do not push the pins in vertically but slant them outwards so they will not obscure the view. Continue pinning the wall for about one inch and cutting the partitions to the fourth somite. Remove the pin at the fourth somite. Cut and pin the rest of the body wall as previously described. Do not damage the brain which is just under the body wall of the **first segment** (Figure 15.2, Plate 15.2).

The digestive system includes a **mouth** and muscular **pharynx**. A relatively long (Segments 6-13) **esophagus** connects the pharynx to the storage **crop** and muscular **gizzard**. Using the probe, push on both bag-like structures and determine their texture. Trace the **intestine** down to the **anus**. Make a transverse cut of the intestine. Cut a section about three mm. in length. Wash the matter out with water. Notice the dorsal fold within the intestine called the **typhlosole** (Figure 15.3, Plate 15.3).

At the head region, search for five dark tubular hearts or aortic arches. These branch off the dorsal blood vessel and cover the esophagus area. Blood flow moves forward in the dorsal blood vessel into the arches and continues back to the tail in the ventral blood vessel (The ventral blood vessel is obscured by the various organs).

Look closely in any coelomic compartment—except the first three and last one. Each kidney or **nephridia** is a coiled, whitish-colored tube. With the aid of a hand lens or dissecting microscope, find a number of nephridia. Notice that each compartment contains a pair of nephridia.

The earthworm is monoecious, having both male and female reproductive organs. Identify three pairs of large, light **seminal vesicles** and two pairs of "beady" **seminal receptacles**. Other sexual parts are too small to identify in this dissection. To find the **circumpharyngeal** and **suprapharyngeal** ganglia (brain), carefully dissect away the pharynx. The small clusters of ganglia can be seen just on either side of the pharynx.

Obtain a prepared slide of the cross-section of **Lumbricus**. Identify: the **cuticle** covering the external **epidermis** and the **muscle layer** beneath it. Regions of both **circular** and **longitudinal** muscles can be differentiated. Find the rounded **intestine** with the surrounding **chloragogue** cells. Pay close attention to the infolded wall of the intestine or **typhlosole**. The **dorsal** and **ventral blood vessels** can be located attached above and below the intestine. Near the muscle wall of the ventral surface is also the circular tube or **ventral nerve cord**. Beneath the muscle layer and surrounding the intestine lies the true body cavity or coelom. If the cross cut section has cut through them, the paired ventral and lateral setae may be visible. Paired **nephridia** will appear as coiled tubules on either side of the intestine in the coelomic cavity.

Class Polychaeta

Bristleworms

On the demonstration table are specimens of polychaetes such as the sandworm, *Nereis* or *Neanthes*. Its segmented body has lateral extensions with bristles or **parapodia**. *Arenicola*, or lugworm, burrows in muddy soil along coastal areas. *Chaetopterus*, the parchment tube worm, produces a U-shaped tube. Specialized parapodia flatten out into fan-like wings directing a water current through the tube and across the body (Figure 15.4, Plate 15.4).

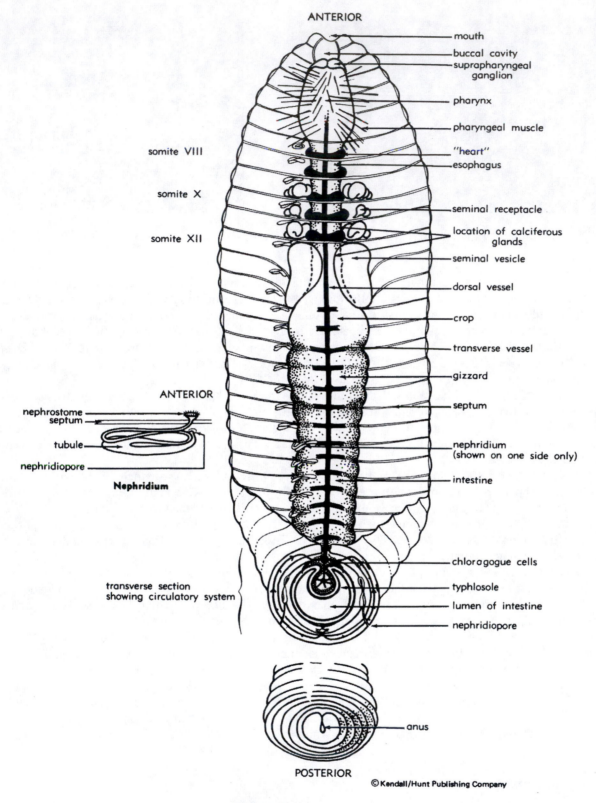

ANTERIOR

mouth
buccal cavity
suprapharyngeal ganglion
pharynx
pharyngeal muscle
"heart"
esophagus
seminal receptacle
location of calciferous glands
seminal vesicle
dorsal vessel
crop
transverse vessel
gizzard
septum
nephridium (shown on one side only)
intestine
chloragogue cells
typhlosole
lumen of intestine
nephridiopore

somite VIII
somite X
somite XII

ANTERIOR

nephrostome
septum
tubule
nephridiopore

Nephridium

transverse section showing circulatory system

anus

POSTERIOR

© Kendall/Hunt Publishing Company

LUMBRICUS

FIGURE 15.2   Earthworm (Internal)

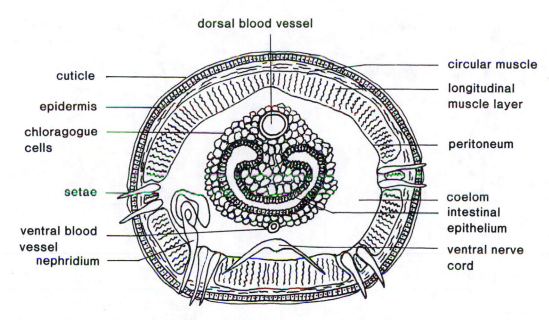

FIGURE 15.3 Earthworm (Cross Section)

PLATE 15.3 Earthworm (Cross Section)

Class Hirudinea

Leeches

Leeches are annelids that have anterior and posterior suckers for locomotion and attachment. They may be aquatic, marine, or terrestrial. They resemble most annelids except that they lack the characteristic setae or parapodia and have a set number of segments. Observe specimens of the medicinal leech, *Hirudo medicinalis.* The blood sucking parasites have an anticoagulant in their mouth secretions and they are still used today to remove accumulated blood from bruised skin areas (Figure 15.4, Plate 15.5).

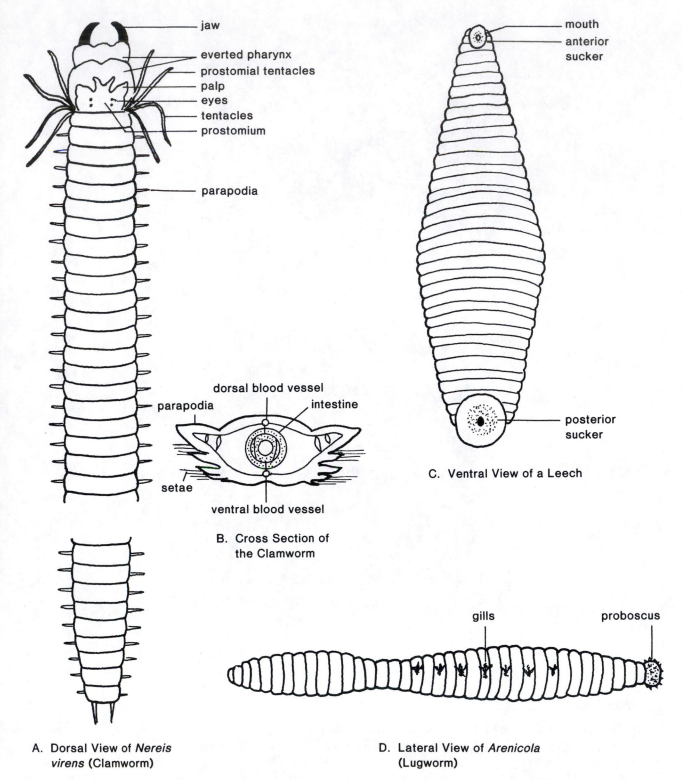

jaw

everted pharynx

prostomial tentacles

palp

eyes

tentacles

prostomium

parapodia

dorsal blood vessel

intestine

parapodia

setae

ventral blood vessel

B. Cross Section of the Clamworm

mouth

anterior sucker

posterior sucker

C. Ventral View of a Leech

gills

proboscus

A. Dorsal View of *Nereis virens* (Clamworm)

D. Lateral View of *Arenicola* (Lugworm)

FIGURE 15.4 Representative Annelid Classes

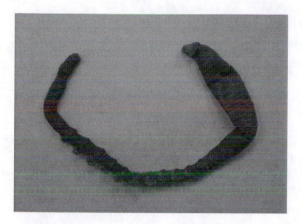

PLATE 15.4   Polychaete

PLATE 15.5   Leech

Questions

1. What type of symmetry is present in the annelids?

2. What is meant by a closed circulatory system? What is the adaptive advantage of such a system?

3. What substances are removed by the nephridia?

4. Explain the differences between seminal receptacles and seminal vesicles.

5. Without lungs or gills, how is respiration accomplished in the earthworm?

6. Of what economic significance are annelid members?

EXERCISE 16   Phylum Arthropoda

Animals with Jointed Appendages

Chelicerata and Crustacea

Learning Objectives

✔ Explain how tagmatization in arthropods has been important in the creation of the arthropod body plan.

✔ Explain how the exoskeleton has contributed to the success of the arthropods.

✔ Explain how jointed appendages have contributed to the success of the arthropods.

✔ Explain how the arthropod diet has been determined by the type of mouth appendages.

✔ Identify the four main features of the chelicerates.

✔ Identify the five main features of the crustacea.

Introduction

The phylum Arthropoda is a large and varied group of animals that constitute more than three-fourths of all known species of animals, approximately a million species. Success of the phylum is obvious, whether measured by total numbers, species, adaptability, or structural variety.

Although arthropod bodies are segmented like that of the annelids, they have achieved the most diversity in body form by dramatically changing and adapting to life on land. To reduce drying out, and provide protection and muscle anchorage, arthropods have a rigid **exoskeleton** made of chitin. Although chitin is found in other organisms, no other group uses this material as extensively as arthropods.

Second, the arthropod body has been regionalized into rigid, functional units or **tagmata**. This segregation forms distinctive **head, thorax** and **abdomen** regions. Third, from these body tagmata, specialized and **jointed** appendages cooperate in feeding, reproduction, and dramatic movement. Members of each subphylum and class modify the arthropod form into the diverse numbers present on the planet.

Subphylum Chelicerata

Class Merostomata

Example: *horseshoe crab (Limulus)*

The more primitive of the arthropod phyla include some fascinating members such as the spiders, scorpions, ticks, mites and the horseshoe crab. All of these animals share four common features: (a) no antennae; (b) no mandibles (chewing mouthparts); (c) anterior appendages are modified into **chelicera** and pedipalps; and (d) the body has two tagmata: a fused head/thorax, or **cephalothorax** and the abdomen.

Observe the shell of a horseshoe crab. Note the armored shell of the cephalothorax, the **prosoma**, and the covered abdomen (**opisthosoma**) including a spined tail or **telson**. Also observe the **compound eyes** on the dorsal surface of the prosoma. Turn the animal over and carefully look at the appendages. The first pair are pincered **chelicera** for manipulating food. The second are the longer **pedipalps** and the last four pair are **walking legs**. Horseshoe crabs belong to an ancient group of animal survivors. Early fossils, 500 million years old, are evidence that the body and lifestyle of the merostomata have been very successful (Figure 16.1, Plate 16.1).

Subphylum Chelicerata

Class Arachnida

The general body form of the chelicerates is modified in the arachnids and represented by the spiders, scorpions, ticks, mites. The fused **cephalothorax** and **abdomen** tagmata are retained in most but can be fused into one skeleton, as in mites and ticks. The **chelicera** vary from the small, manipulating appendages of scorpions and ticks to the terminal

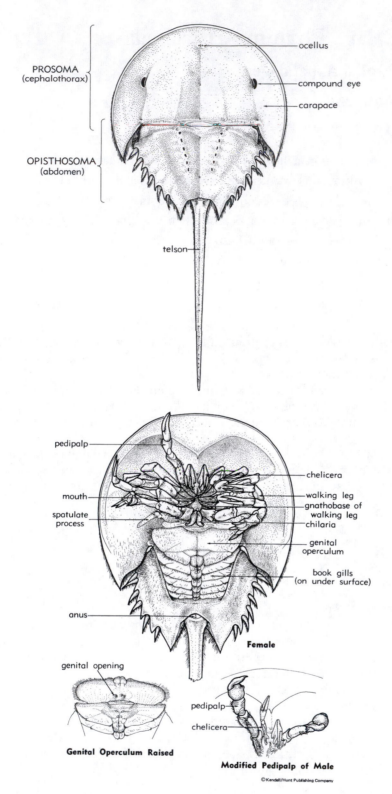

FIGURE 16.1 Horseshoe Crab (Limulus)

PLATE 16.1   Horseshoe Crab

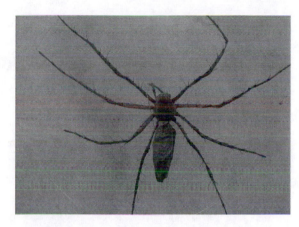

PLATE 16.2 A   Spider

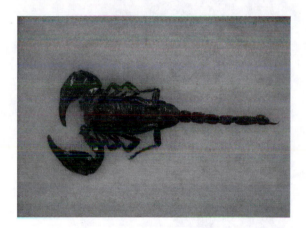

PLATE 16.2 B   Scorpion

poison fangs in spiders. **Pedipalps** are usually food manipulating appendages but are dramatically modified in the scorpion's grasping claws. Four pairs of **walking legs** are also attached to the cephalothorax. The segmented abdomen may have the familiar poison **stinger** of scorpions or the **silk glands** and weaving **spinnerets** of spiders. Observe dried, jarred or table specimens of the arachnid group and identify specific examples and recognizable body parts (Figure 16.2, Plate 16.2 A-B). If prepared slides are available of mites (*Sarcoptes, Demodex*) or ticks (*Dermacentor, Boophilus*), use the dissecting microscope and view their anatomy.

Subphylum Crustacea

Class Malacostraca

Example: *Crayfish (Astracus or Cambarus)*

Members of the subphylum Crustacea include the crayfish, crab, shrimp, water flea, barnacle, and other types. The great majority of crustaceans are aquatic or marine, while a few live on land. Crustaceans possess all of the characteristics of arthropods, but are distinguished from other classes of arthropods by having two pairs of **antennae**, two tagmata, the cephalothorax and abdomen, one pair of chewing **mandibles** and two pairs of oral **maxillae**. The appendages of crustacea are forked at the terminal end and are called **biramous**. Since most crustaceans are aquatic, respiration is by **gills** protected by the exoskeleton.

External Anatomy

Obtain a preserved crayfish, hand lens, and a dissecting pan. Place the crayfish in the pan in its usual upright position. Starting at the anterior end, identify the various parts. The **cephalothorax** is covered by a continuous **carapace**

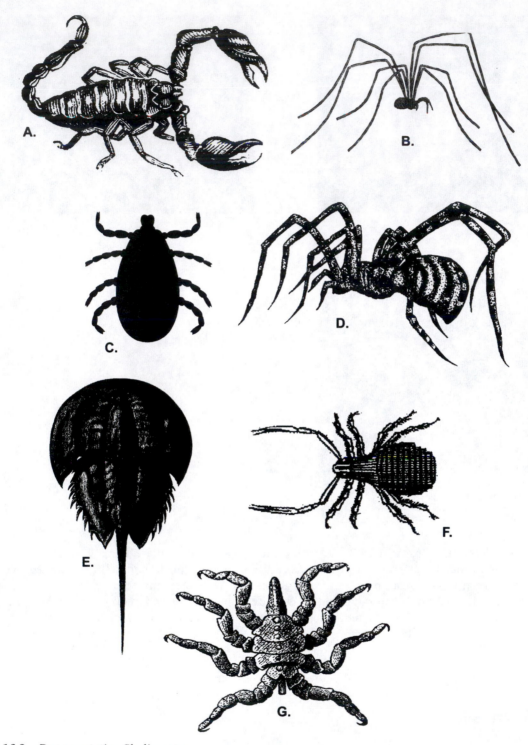

FIGURE 16.2   Representative Chelicerates

which acts as a protective shield. A **cervical groove** marks the division between the head and the thorax. At the anterior end of the carapace is a median pointed **rostrum**. On either side beneath the rostrum is a stalked, movable **compound eye.** Locate the two pairs of antennae: the smaller **antennules** and the longer **antennae.** The abdomen ends in a flared tail segment with one central **telson** and winged **uropods** (Figures 16.3, 16.4, 16.5, Plates 16.3 A-B, 16.4 A-B).

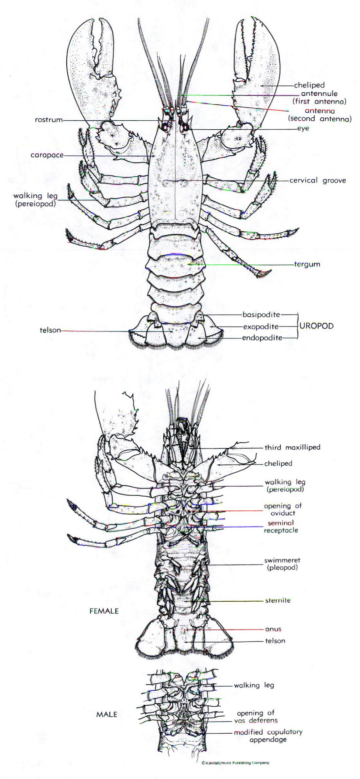

FIGURE 16.3   Crayfish (External)

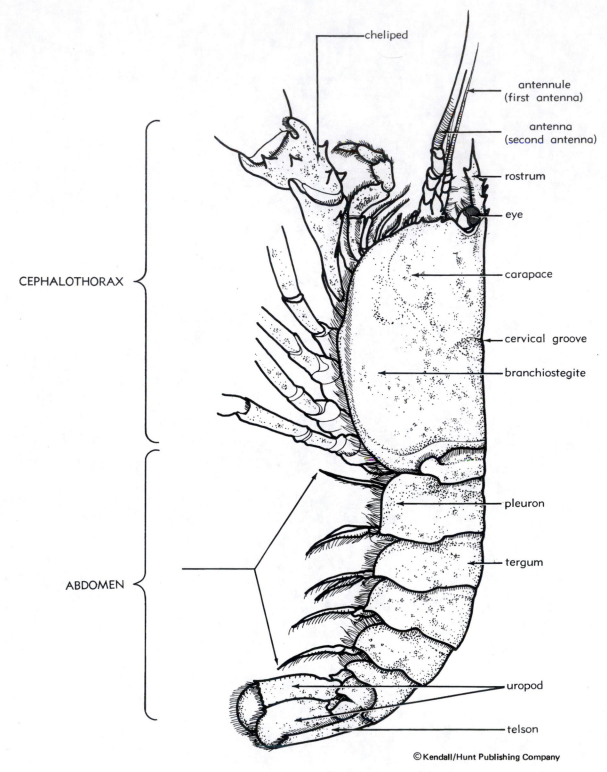

FIGURE 16.4   Crayfish (Lateral)

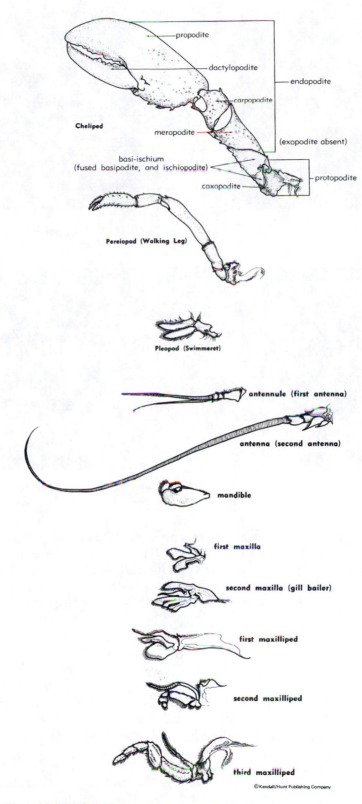

FIGURE 16.5   Crayfish (Varied Appendages)

Turn the crayfish over so the ventral surface is now visible. Starting at the mouth fold back and locate the **mandibles**, two **maxillae** and three **maxillipeds** (respectively) around the mouth. The first walking leg pair is modified

PLATE 16.3 A   Crayfish (Dorsal)

PLATE 16.3 B   Crayfish (Ventral)

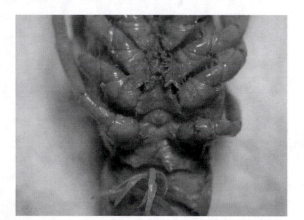

PLATE 16.4 A   Crayfish (Female)

PLATE 16.4 B   Crayfish (Male)

into the **cheliped**. Four other pairs of **walking legs** or **pereiopods** are used in locomotion and food manipulation. The abdominal appendages or **swimmerets** create water currents and help to carry the eggs on female bodies. Sexual differences can be seen in the first two pairs of swimmerets. In the male, the larger, stiffer, **gonopods** are modified for sperm transfer. In the female, the first two pairs of swimmerets are nondescript or greatly reduced and the opening of the **oviduct** is seen between walking legs #3 and #5. Finally, note the **anus** opening in the **telson** segment.

Internal Anatomy

Carefully remove the carapace by making an initial cut at the cephalothorax and abdomen joint and snipping along the mid-dorsal line **forward** to a point just behind the eyes. Leave the eyes in place. Gently free up the front edge of the carapace and separate the soft tissue away from the exoskeleton. Remove both sides of the carapace. The soft, feathery **gills** cover both sides. Directly behind the eyes, locate the double pouched **stomach** (**cardiac and pyloric**) with a curved **gastric mill** forming a rigid ring. On either side should be a light mass of tissue, the **digestive glands** that aid in the digestive process. The shield-like **heart** is centrally located behind the stomach. Posterior and lateral to the heart are the paired **gonads**.

Gently remove the stomach, heart and digestive glands and locate two circular, plate-like **green glands** on either side of the mouth opening. These antennal or maxillary glands are used for osmoregulation. Two white, thin **nerve tracts** point anteriorly to the larger ganglion (**brain**) (Figures 16.6, 16.7, Plate 16.5).

Pointing the scissors posteriorly, now continue a center cut through the abdominal shell and peel it away. Identify the central tube, **intestine**, and the large flexor and extensor **muscles**.

Other Crustacea

Examine other types of crustaceans on the demonstration table (crab, shrimp, barnacles, etc.). Turn over each sample and identify the characteristics that places them in the crustacean group.

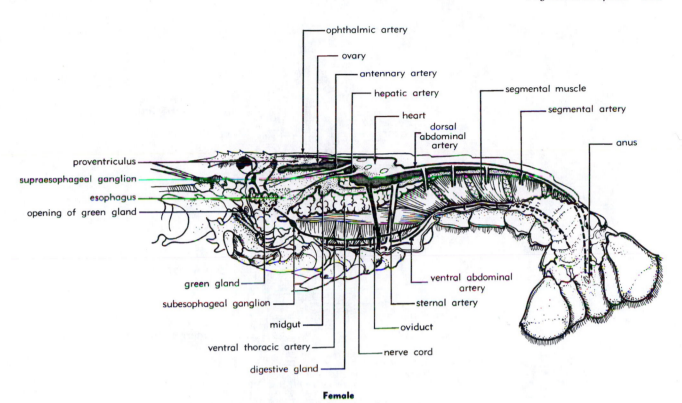

ophthalmic artery
ovary
antennary artery
hepatic artery
heart
dorsal abdominal artery
segmental muscle
segmental artery
anus
proventriculus
supraesophageal ganglion
esophagus
opening of green gland
green gland
subesophageal ganglion
midgut
ventral thoracic artery
digestive gland
ventral abdominal artery
sternal artery
oviduct
nerve cord

Female

LEFT DIGESTIVE GLAND REMOVED; TAIL TURNED TO SHOW VENTRAL SURFACE

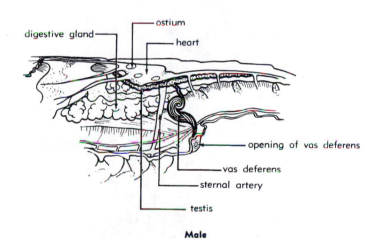

digestive gland
ostium
heart
opening of vas deferens
vas deferens
sternal artery
testis

Male

AREA OF GONADS; LEFT DIGESTIVE GLAND REMOVED

FIGURE 16.6 Crayfish (Dissection)

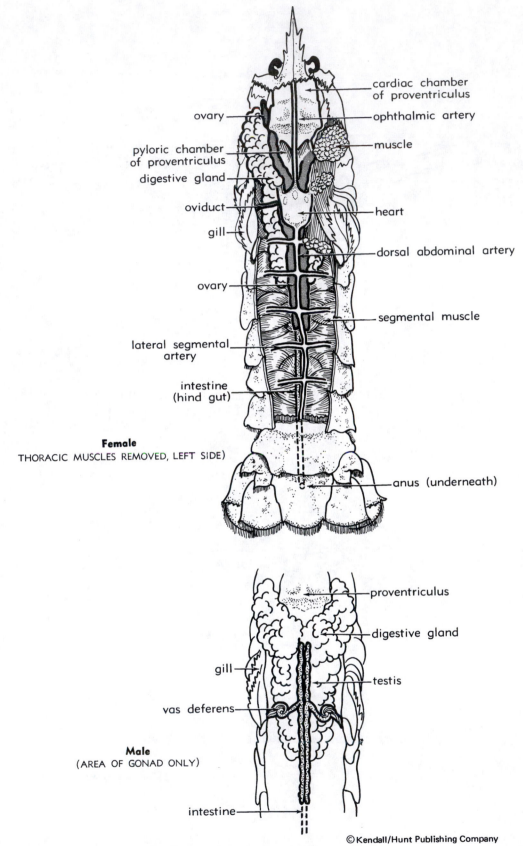

cardiac chamber
of proventriculus

ovary

ophthalmic artery

pyloric chamber
of proventriculus

muscle

digestive gland

oviduct

heart

gill

dorsal abdominal artery

ovary

segmental muscle

lateral segmental
artery

intestine
(hind gut)

Female
THORACIC MUSCLES REMOVED, LEFT SIDE)

anus (underneath)

proventriculus

digestive gland

gill

testis

vas deferens

Male
(AREA OF GONAD ONLY)

intestine

© Kendall/Hunt Publishing Company

FIGURE 16.7   Crayfish Dissection (Dorsal View)

PLATE 16.5   Crayfish (Internal)

Questions

1. Name some crustaceans that might be found in a terrestrial habitat.

2. What sensory organs does the crayfish have that inform it about its surroundings? Which of these would be helpful in muddy water?

3. What is the function of the different types of crayfish appendages?

4. Of what medical and economic significance are the arachnids?

5. Of what economic significance are the crustaceans?

EXERCISE 17 Phylum Arthropoda

Animals with Jointed Appendages

Subphylum Uniramia

Learning Objectives:

- ✔ Identify the physical characteristics of insects that separate them from the other arthropods.
- ✔ Describe how insects have solved the problem of survival on dry land.
- ✔ Recognize members of various insect orders.
- ✔ Identify the positive and negative impact that insects have on the planet.

Introduction

The Uniramia are identified as arthropods whose appendages have only one (uni) branch. They possess the basic characteristics of exoskeleton, jointed appendages and tagmata that have been seen in the chelicerata and crustacea. The insects have exploded in numbers and varieties by adapting well to terrestrial life. They have improved on the basic plan with well developed sensory organs, unparalleled reproductive fecundity and varied their mouthparts to exploit all types of food. Furthermore, they have complicated behavior patterns, developed social orders, undergo metamorphosis and most dramatically achieved flight.

Class Chilopoda

The centipedes are land arthropods with flattened bodies. The head appendages are similar to the insects and their somite number can range from a few to over 100. Each segment, excluding the first and last two, bear a pair of jointed legs. The first pair of these "hundred-leggers" are modified into poison claws. The centipedes search for prey in moist places, such as under logs and stones, paralyze them with the poison claws and chew them with mandibles. Observe preserved centipedes from the display table. Using a dissecting scope, focus on the anterior end and look for the modified head appendages (Plate 17.1).

Class Diplopoda

Millipedes with 11 to 100 segments differ from the centipedes by having two pairs of appendages on each somite. More rounded in body form, their "thousand-legs" propel them quickly yet strongly through ground litter to feed on plant matter. Many of these herbivores also possess **repugnatorial glands** capable of spraying hydrogen cyanide when threatened. Compare preserved millipedes to centipedes and be prepared to tell them apart (Plate 17.2).

PLATE 17.1   Centipede

PLATE 17.2   Millipede

Class Insecta

Example: *Romalea microptera*

External Anatomy

Obtain a preserved grasshopper, hand lens and a dissecting pan. Place the grasshopper in the pan in its usual upright position. Starting at the anterior end, identify the various parts. Note the body is divided into three tagmata: head, thorax and abdomen. On the head tagmata, locate the single pair of **antennae**, one pair of **compound** eyes, three simple eyes or **ocelli** and the mouth parts. Remove each mouth part by grasping the piece at its base with forceps and pulling sharply. These include the upper **labrum** with its base clypeus, two hard, serrated **mandibles**, two **maxillae** with palps and a lower **labium**. The hypopharynx remains as the last piece in the center of the mouth (Figure 17.1, 17.2, 17.3).

The thorax is composed of three fused somites. On the first, **prothorax**, is attached a pair of **walking legs**. A second pair of walking legs and a pair of **forewings** or protective wings are attached to the mesothorax. Finally a pair of jumping legs and a pair of flying or **hindwings** are attached to the **metathorax**.

Note the abdominal segments. The first one has an oval window or **tympanum** that functions in hearing. The last three segments are modified for reproductive purposes. In the female spiny **ovipositors** are utilized for egg deposit; in the male tail segments are modified for copulation. Along either side of the abdomen, locate very small openings or **spiracles** which are external openings for respiration.

Identify the individual leg parts. Starting from the attachment point, the **coxa** and **trochanter** form a rotating cuff and articulates with the **femur**. The **tibia** is spiny and terminates in the flexible and clawed **tarsus**. Although the legs have similar parts, the hindleg is larger and obviously modified for jumping.

Internal Anatomy

Remove the legs and wings of the grasshopper. Make a shallow incision in the tip of the abdomen and cut along the mid-dorsal line to just behind the head. Peel away the exoskeleton, being careful to tease away the body tissues. Just anterior to the head, locate the **crop**, used for food storage. Finger-like projections or the **gastric caeca** of the stomach increase the surface area for food absorption. Along the dorsal surface, under a thin membrane, lies the slightly enlarged **heart** and slender, tubular, dorsal **aorta**. **Malpighian tubules** or excretory organs are thin, filamentous projections closer to the tail. In females, oblong **eggs in ovaries** are tightly packed along the dorsal surface of the digestive system. The **oviducts** connect the ovaries to the **vagina** opening. In males, a pair of **testes** are bound together above the digestive tract. Beneath them are the two **sperm ducts** (vas deferens) that meet to form a single **ejaculatory duct** (Figure 17.4 A, B, C).

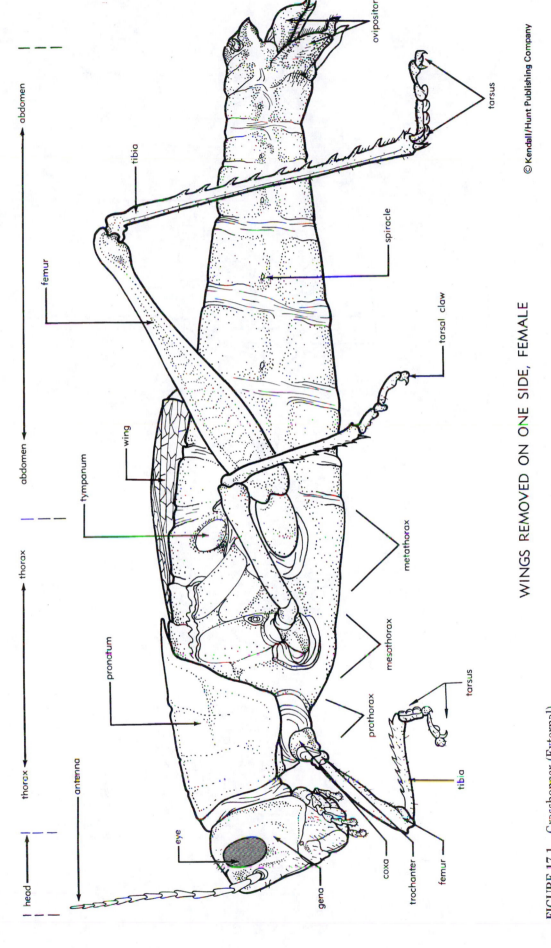

WINGS REMOVED ON ONE SIDE, FEMALE

FIGURE 17.1   Grasshopper (External)

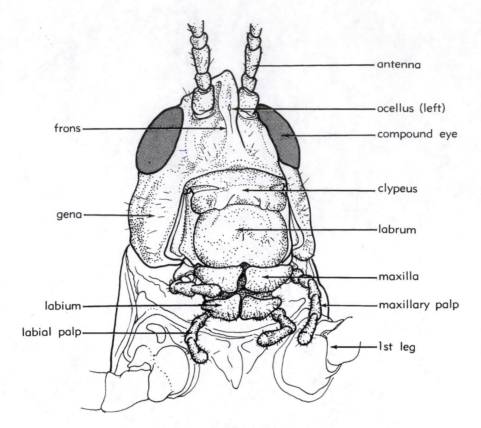

Head, Ventral View

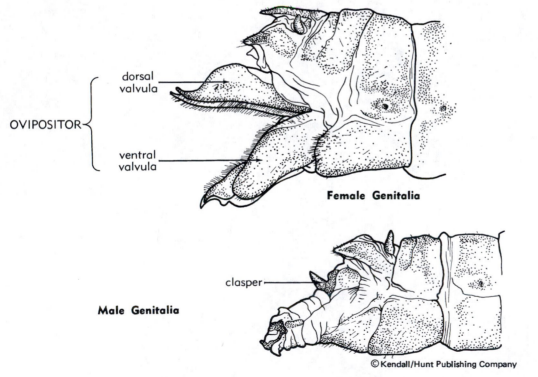

Female Genitalia

Male Genitalia

© Kendall/Hunt Publishing Company

FIGURE 17.2   Grasshopper (External detail)

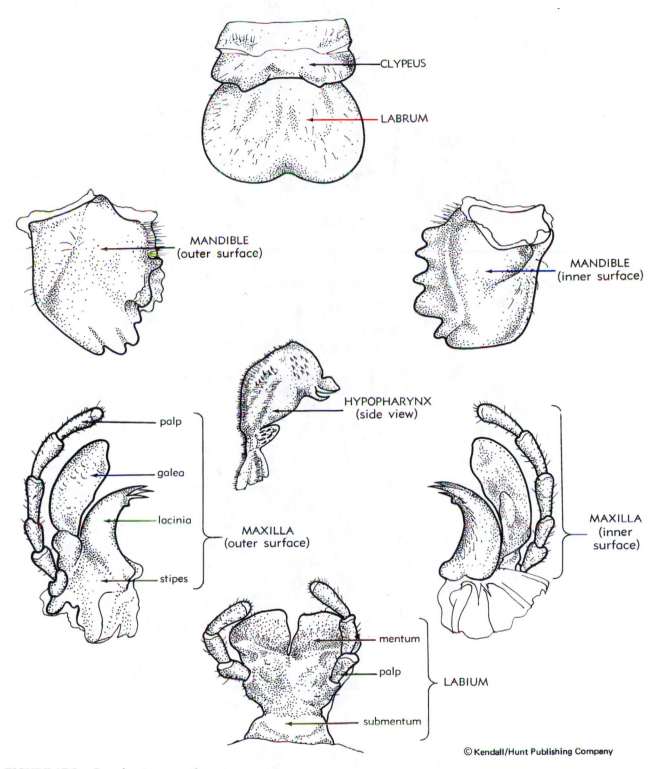

CLYPEUS

LABRUM

MANDIBLE
(outer surface)

MANDIBLE
(inner surface)

palp

galea

lacinia

stipes

HYPOPHARYNX
(side view)

MAXILLA
(outer surface)

MAXILLA
(inner surface)

mentum

palp

submentum

LABIUM

© Kendall/Hunt Publishing Company

FIGURE 17.3 Grasshopper mouth parts

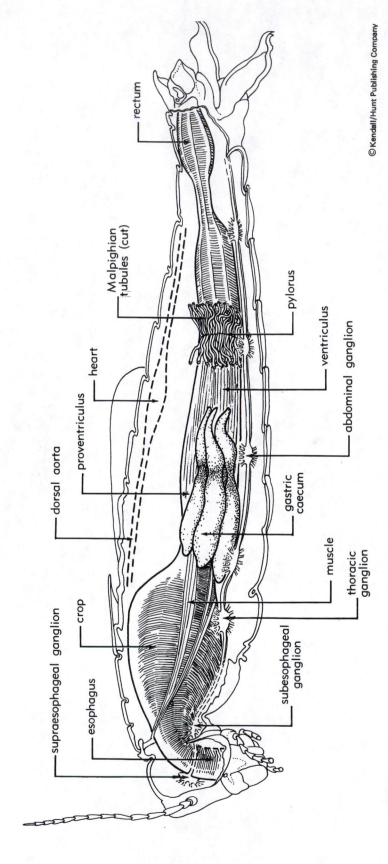

FIGURE 17.4 A Grasshopper (Internal)

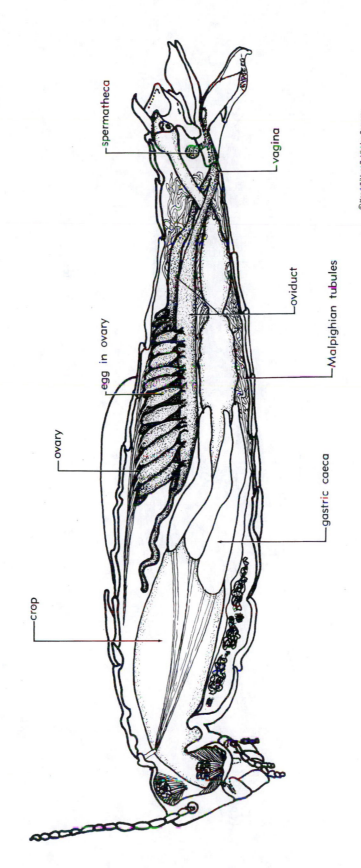

spermatheca

vagina

oviduct

Malpighian tubules

egg in ovary

ovary

gastric caeca

crop

FIGURE 17.4 B Female

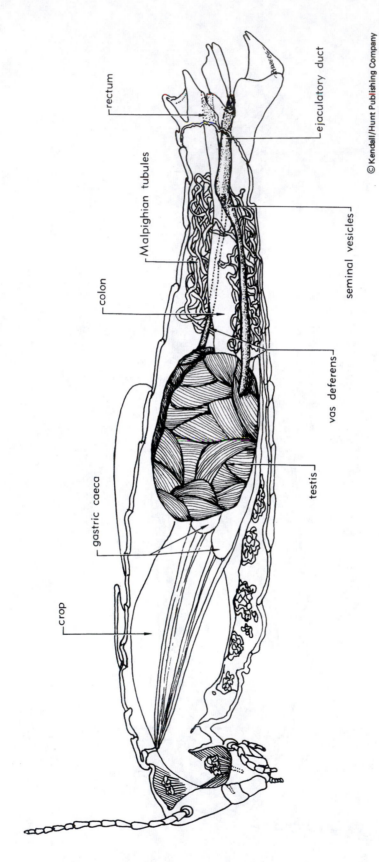

rectum

Malpighian tubules

colon

gastric caeca

crop

testis

vas deferens

seminal vesicles

ejaculatory duct

© Kendall/Hunt Publishing Company

FIGURE 17.4 C Male

Questions

1. What is the difference between compound and simple eyes?

2. How do the two pairs of wings differ in the insect orders?

3. Give at least 4 characteristics of insects that have made them so successful in a terrestrial environment.

4. In what ways do millipedes and centipedes differ from each other?

5. What positive impact do the insects have on the planet?

6. What negative impact do the insects have on the planet?

Worksheet to Distinguish Uniramia and Insect Orders

The following worksheet identifies and describes selected examples of the Subphylum Uniramia with focus on the insect orders. Collect examples of various insects or use preserved specimens to determine their names and orders. Use the dichotomous key to complete this project.

Phylum Arthropoda

The Uniramians

Subphylum Uniramia

Class Chilopoda *(centipedes)*
Examine those centipedes on display. Note the flattened trunk with *a pair of appendages for each segment*. Also observe the large **fangs** located on the maxillipeds just ventral to the elongated antennae. Finally, examine the sensory **cerci** projecting from the final segment.

Class Diplopoda *(millipedes)*
Compare the basic structure of a millipede with that of a centipede. Note the rounded shape, lack of poison fangs, and presence of *two pairs of appendages* for each body diplosegment.

Class Insecta *(insects)*
There are a number of insects on display. Examine each to learn its specific properties. Note the three body regions (**head, thorax,** and **abdomen**), *three pairs of legs* borne on the thorax, and wings (if present). Use the key provided to identify each of the orders portrayed below. Using a dissecting microscope, or a magnifying glass, compare each to the description given.

Representative Insect Orders

Ephemeroptera *(mayflies)*
Mayflies are primitive insects with a prominent aquatic larval stage. The adult part of life cycle is reduced; adults live for only a few days, at most, during which time they don't feed. They mate in swarms while flying. When at rest, the wings are held vertically over the back; the abdomen has three long thread-like cerci, legs are usually small, mouthparts are pronounced in larva but vestigial in adults. Mayflies are an important food source for many fish.

Odonata *(dragonflies and damselflies)*
Aquatic larvae are predacious, as are the adults, which prey on flying insects. These primitive insects have large compound eyes, reduced, bristlelike antennae, long and slender abdomens, chewing mouthparts, and two pairs of net-veined wings held out to the side when at rest (dragonflies) or vertically over the back (damselflies).

Orthoptera *(grasshoppers, cockroaches, katydids, crickets, walking sticks, mantids)*
Legs adapted for running (cockroaches), jumping (grasshoppers, crickets, katydids), or raptorial (mantids). Forewings narrow and leathery, hind wings usually large and fan-like, chewing mouthparts, cerci present, antennae moderate to very long, female with pronounced terminal ovipositor, sound, reception is a tympanum on legs or abdomen, and parts of the exoskeleton may stridulate to produce sound.

Isoptera *(termites)*

All termites are eusocial with a well-defined caste system of blind workers, blind soldiers with enlarged mandibles, and reproductives (males are winged but live for only a brief period, queens are initially winged, but shed them to become large reproductive units cared for by workers). These are small, soft-bodied insects. The abdomen is broadly joined to thorax, they live in huge swarming colonies, highly destructive when they occur in human residences.

termite queen

Dermaptera *(earwigs)*

In some locations, earwigs are common household residents, but they are scavengers and present no health threat. They possess enlarged forceplike abdominal cerci used in defense, hind wings are large and folded under short, horny forewings.

Anoplura *(sucking lice)*

These are blood-sucking ectoparasites of birds and mammals, including humans. Head lice (cooties) and pubic lice (crabs) have infested mankind throughout history. They have piercing-sucking mouthparts, wings are absent, legs adapted for attachment, compound eyes missing, broad body capable of great distention during a blood meal.

Hemiptera *(true bugs)*

Piercing-sucking mouthparts form a jointed needlelike beak, that arises from the front portion of the head and is held between the forelegs when not in use. Both sets of wings lie flat against the abdomen when at rest; forewings thick and leathery at the base and membranous toward the tip, hind wings large and membranous but folded beneath the forewings. Prominent triangular plate (the scutellum) located between the bases of the forewings. Most are herbivores, many are predacious, some are ectoparasites (kissing bugs and bed bugs both infest humans). As a group, hemipterans are of great economic importance as destroyers of crops.

Homoptera *(cicadas, leafhoppers, plant hoppers, aphids)*

Feed on plant sap, mouthparts modified for piercing and sucking. Some forms are wingless but most have two pair of membranous wings held tentlike or rooflike over the abdomen when at rest.

Hymenoptera *(ants, bees, and wasps)*

This order is characterized by the presence of a narrow waist that joins the abdomen to the thorax. Chewing mouthparts, some have a stinger derived from the ovipositor, hind wings small and joined to larger forewings by a hook. Tendency in this group to form eusocial units. Some species are nuisances, others are highly beneficial in pollinating plants.

Siphonoptera *(fleas)*

Small, wingless insects, body laterally compressed, mouthparts modified to feed exclusively on blood, hindlegs adapted for jumping, other legs for clinging. All are ectoparasites and some are transmitters of serious disease (i.e., bubonic plague). Adults are not highly specific and will switch from one host species to another; humans are often infested with dog, cat, or rat fleas, as well as their own.

Diptera *(flies, mosquitoes, and gnats)*

One pair of wings is usually present, the other pair reduced to knobby structures called halteres used as gyroscopes for balance while flying, large eyes, highly mobile head. Mouthparts are variable and may be adapted for sponging, piercing, or chewing. Of great economic and medical importance; nuisance Lepidoptera (butterflies and moths). Large compound eyes with long antennae, wings and body covered with overlapping, dense, pigmented scales. Mouthparts modified into a long, coiled tube to extract nectar. Butterflies hold wings vertically at rest, moth, horizontally. Larvae of great economic importance as destroyers of crops and cloth; adults are important pollinators.

Coleoptera *(beetles)*

This is the largest order of insects; there are more beetles than all non-arthropods combined. Head bears well-developed antennae and eyes, two pairs of wings with the first pair consisting of a hard shell-like covering, the elytra, that meets in a straight line down the back, hind wings membranous and folded under forewings when at rest, body compact and hard, mouthparts chewing. Some are serious agricultural pests, while others are predators of harmful pests.

Key of Selected Insect Orders*

1. a. Wings present and well developed 2
   b. Wings absent 20
2. a. Forewings thickened or leathery (at least at base); hind wings membranous, may be hidden beneath forewings 3
   b. All wings membranous throughout 7
3. a. Mouthparts beaklike (somewhat like a syringe and often held between legs when not in use) 4
   b. Mouthparts mandibulate, adapted for chewing 5
4. a. Beak arises from front of head, forewings leathery at base but membranous at tips (true bugs) . . . . . . Hemiptera
   b. Beak arising from rear of head, forewings uniform and held tentlike over abdomen (plant hoppers) . . . . . Homoptera
5. a. Abdomen with forcepslike cerci, forewings short and covering folded hind wings (earwigs) . . . . . **Dermaptera**
   b. Not as above 6
6. a. Forewings hard, veinless, shell-like meeting in a straight line; hind wings folded under forewings (beetles) . . . . . **Coleoptera**
   b. Forewings not as above, veined, hind wings broad and usually shorter than forewings; forelegs may be raptorial, hind legs may be modified for jumping (grasshoppers, mantids, crickets, cockroaches) . . . . . **Orthoptera**
7. a. One pair of wings only 8
   b. Two pair of wings 11
8. a. Body grasshopper-like, hind legs adapted for jumping . . . . . **Orthoptera**
   b. Not as above 9
9. a. Mouthparts vestigial (missing); three long, threadlike cerci (earwigs) . . . . . **Dermaptera**
   b. Mouthparts chewing, no cerci 10
10. a. Sucking mouthparts; hind wings reduced to clublike halteres, tarsi five segmented (flies) . . . . . **Diptera**
    b. Mouthparts chewing, tarsi two or three segmented . . . . . **Psocoptera**
11. a. Wings covered by overlapping, pigmented scales; mouthparts coiled (butterflies/moths) . . . . . **Lepidoptera**
    b. Not as above 12
12. a. Wings long and narrow, fringed with extremely long hairs (edges appear fuzzy) . . . . . **Thysanoptera**
    b. Not as above 13
13. a. Forewings large and triangular; hind wings small and rounded; wings held vertically over back; wings heavily veined; soft-bodied insects with two or three long, threadlike tails (mayflies) . . . . . **Ephemeroptera**
    b. Not as above 14
14. a. Tarsi five segmented (just count the segments on the distal leg region [tarsus]) 15
    b. Tarsi consisting of four or fewer segments 16
15. a. Rather hard-bodied, wasplike insects; usually a narrow waist attaches abdomen to thorax; hind wing smaller than forewing . . . . . (bees, wasps, yellowjackets) . . . . . **Hymenoptera**
    b. Not as above 16
16. a. Hind wings and forewings of equal size; wings with many veins; antennae short and bristlelike; abdomen long and slender; wings outstretched at rest (dragonflies) or held vertically (damselflies) . . . . . **Odonata**
    b. Not as above 17
17. a. Sucking mouthparts (mouthparts basically a tube or stylet) 18
    b. Chewing mouthparts 19

*Modified from Borrer and DeLong, An Introduction to the Study of Insects. Holt, Rinehart & Winston.

18. a. Beak arising from front of head (true bugs) . . . . . **Hemiptera**

    b. Beak arising from hind part of head (cicadas, aphids) . . . . . **Homoptera**

19. a. Wings similar to each other; soft bodied; cerci small, winged termites . . . . . **Isoptera**

    b. Not as above 20

20. a. Ectoparasites of birds and mammals; body flattened laterally or dorso-ventrally 21

    b. Not as above; soft bodied, usually blind insects, may have large pincerlike mandibles (wingless termites) . . . . . **Isoptera**

21. a. Body flattened laterally; jumping hind legs; body sparsely covered with spine or hairs (fleas) . . . . . **Siphonaptera**

    b. Body flattened dorso-ventrally; large tarsal claws for attachment to host (sucking lice) . . . . . **Anoplura**

EXERCISE 18 Phylum Echinodermata

Spiny-Skin Animals

Water Vascular System

Learning Objectives

✔ Describe the general structure of the echinoderm body plan.

✔ Describe the structure and function of the water vascular system.

✔ Recognize representative specimens of the various classes of echinoderms.

Introduction

The phylum Echinodermata includes sea stars, sea urchins, brittle stars, sea cucumbers, and feather stars. They form a large group of specialized marine organisms that are interesting because of their bizarre body form and their evolutionary relationship to the chordates. The echinoderm larvae undergo a striking change during their embryonic development from a form with bilateral symmetry to one with radial symmetry. The resulting symmetry, *pentaradiate* symmetry, always has the body parts arranged radially in fives or multiples of five. Apparently this type of symmetry suits the aquatic environment but reduces the need for major organ systems. Thus, they lack gills or lungs, a brain, a heart, kidneys, a distinctive head region and rely on external fertilization.

A unique characteristic of echinoderms is the water vascular system, that consists of numerous water-filled tubes ending in a large number of tube feet. They have: a coelomic cavity lined with ciliated peritoneum, an endoskeleton of calcareous plates, a spiny epidermis, dermal papillae as respiratory organs and a simple nervous system.

The echinoderms are deuterostomes like the higher organisms; namely the chaetognatha, hemichordates and the chordates. The embryology of sea stars was observed in Exercise 6. Review the development of the gastrula and determine why the echinoderms are placed in the deuterostome line, linking them in a common origin to the higher organisms.

Class Asteroidea

Example: *Asterias*

External Anatomy

This class includes sea stars, or more commonly called starfish. The lab specimens have five rays or arms; however, some species may have six, seven, eleven, or up to 50 in number. Obtain a preserved sea star and place in a dissecting pan. Make a distinction between the oral (ventral) and aboral (dorsal) surfaces (Figures 18.1, 18.2, 18.3, Plate 18.1). On the aboral surface locate the button-like structure, madreporite, on the central portion called the disc. The disc has five arms or *rays* attached. The madreporite is between two arms, called the bivium, while the remaining arms are called the trivium.

With the aid of a hand lens or dissecting microscope, find little projections which form a ring around the base of each spine. These are called pedicellariae, pincher-like structures which keep the surface area free from debris. Also, distinguish soft, finger-like projections between the spines, these are the dermal papillae.

Internal Anatomy

To dissect the specimen, cut off the end of a trivium about one-half inch from the tip with a pair of scissors. Then make two lateral cuts extending toward the disc. Remove the body wall from the dorsal surface of the disc by cutting around its margins just inside the arms. Cut around the madreporite, leaving it in place on the disc. Gently lift up the dorsal wall of the arm and disc and gently scrape the attached organs into the bottom portion of each arm. The greenish structures that adhere to the arm coverings are the pyloric caeca (also called hepatic caeca or digestive glands) (Figure 18.2 A). The sac-like stomach (cardiac and pyloric portions) sits in the center of disc and is distendable out the mouth area. A short intestine leads to the anal opening on the aboral surface, both of which are difficult to see.

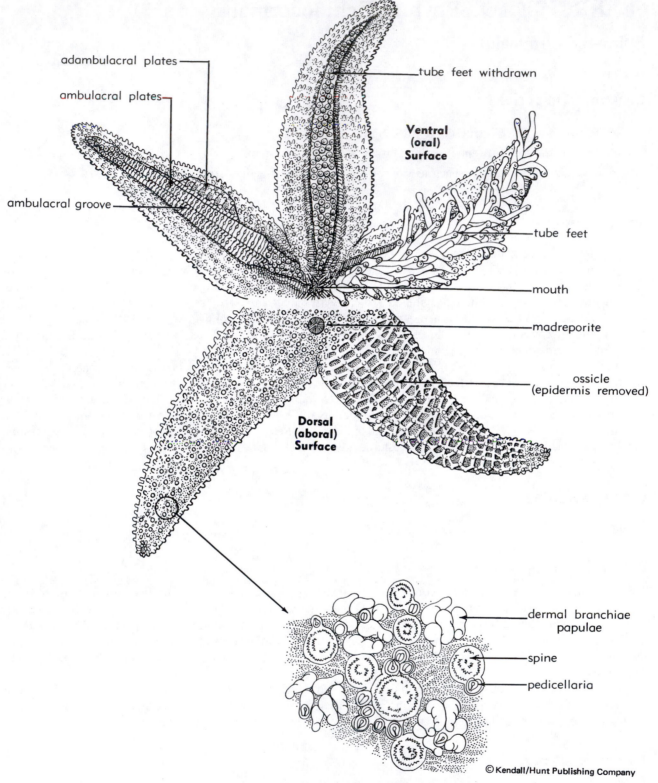

adambulacral plates

ambulacral plates

ambulacral groove

tube feet withdrawn

**Ventral
(oral)
Surface**

tube feet

mouth

madreporite

ossicle
(epidermis removed)

**Dorsal
(aboral)
Surface**

dermal branchiae
papulae

spine

pedicellaria

© Kendall/Hunt Publishing Company

Magnified Area on Dorsal Surface

FIGURE 18.1   Starfish (External)

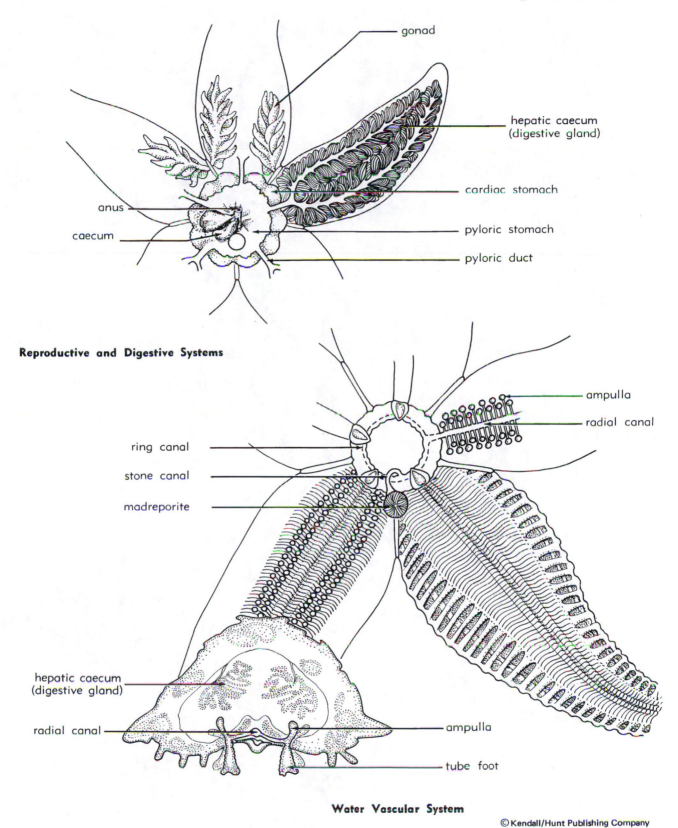

gonad

hepatic caecum
(digestive gland)

cardiac stomach

pyloric stomach

pyloric duct

anus

caecum

Reproductive and Digestive Systems

ampulla

radial canal

ring canal

stone canal

madreporite

hepatic caecum
(digestive gland)

radial canal

ampulla

tube foot

Water Vascular System

© Kendall/Hunt Publishing Company

FIGURE 18.2   Starfish (Internal)

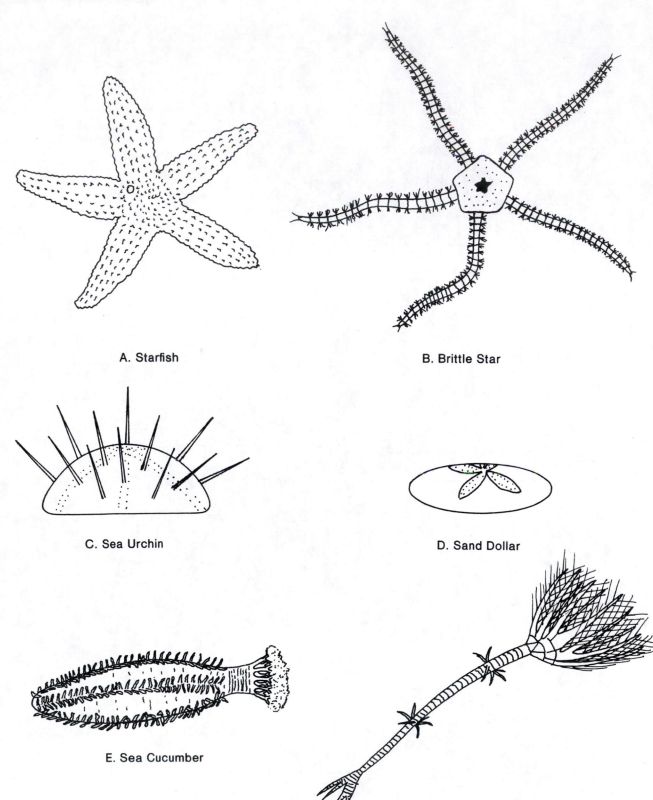

A. Starfish

B. Brittle Star

C. Sea Urchin

D. Sand Dollar

E. Sea Cucumber

F. Sea Lily

FIGURE 18.3   Representative Echinoderm Classes

PLATE 18.1 A   Starfish (aboral surface)

PLATE 18.1 B   Starfish (oral surface)

PLATE 18.1 C Starfish anatomy

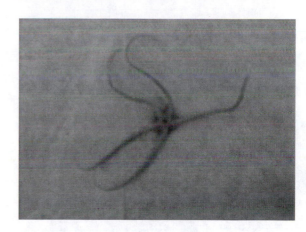

PLATE 2   Brittle Star

A pair of pinkish, glandular **gonads** lie under the pyloric caecae in each arm. Gonads vary in size from very small to those filling the arm's cavity. The sexes are separate; however, to determine the sex, a microscopic examination of the gonadal tissue must be made.

Remove the pyloric caecae and gonads. Note the bulb shaped **ampullae** between the **limy ossicles** (Figure 18.2). Each tube foot is located in the **ambulacral groove**. The ampullae in each arm connect by **lateral canals** to a centrally located **radial canal**. Move to the central disc and remove the pyloric and cardiac stomachs. Be careful not to damage the **stone canal** which leads from the madreporite to the **ring canal** which circles the mouth region.

Class Ophiuroidea

Elongated, spindly arms about a small central disc identify the brittle stars of this class. The extended appendages gives them a graceful, yet wiry locomotion. The tendency of the arms to fragment earns them their common name, while serpent stars identifies their wriggly motion (Figure 18.3, Plate 18.2).

Class Echinoidea

Sea urchins and sand dollars lack rays, but have many movable spines about a hemispherical body. The gonads of some sea urchins are considered to be delicacies in some island cultures. On the oral surface, five pointed teeth form the **Aristotle's lantern** around the mouth. The flattened forms of sand dollars and sea biscuits are covered by downy numerous spines. Once cleaned, their skeletons are collectable items in many beach stores (Figure 18.3, Plate 18.3, Plate 18.4).

PLATE 18.3   Sea Urchin

PLATE 18.4   Sand Dollar

PLATE 18.5   Sea Cucumber

Class Holothuroidea

The class Holothuroidea is represented by the sausage-shaped sea cucumbers. They have a slender elongated body on an oral-aboral axis that is protected by a leathery epidermis. Tentacles around the mouth filter the water for food. Oddly, the sea cucumbers are capable of **evisceration** of body parts when threatened. Hiding away, they will regenerate the soft inner parts left behind (Figure 18.4, Plate 18.5).

Class Crinoidea

The class Crinoidea includes the sea lilies and feather stars. These echinoderms remain attached for much of their lives. Sea lilies have a flower-shaped body at the tip of an attached stalk while the feather stars have long multi-branched arms with no stalk. Crinoid fossils date back to the Cambrian Period, evidence of the lasting power of these fragile looking echinoderms.

Examples of each of the echinoderm classes are available on the display table. Identify the primary features that place them in each respective class.

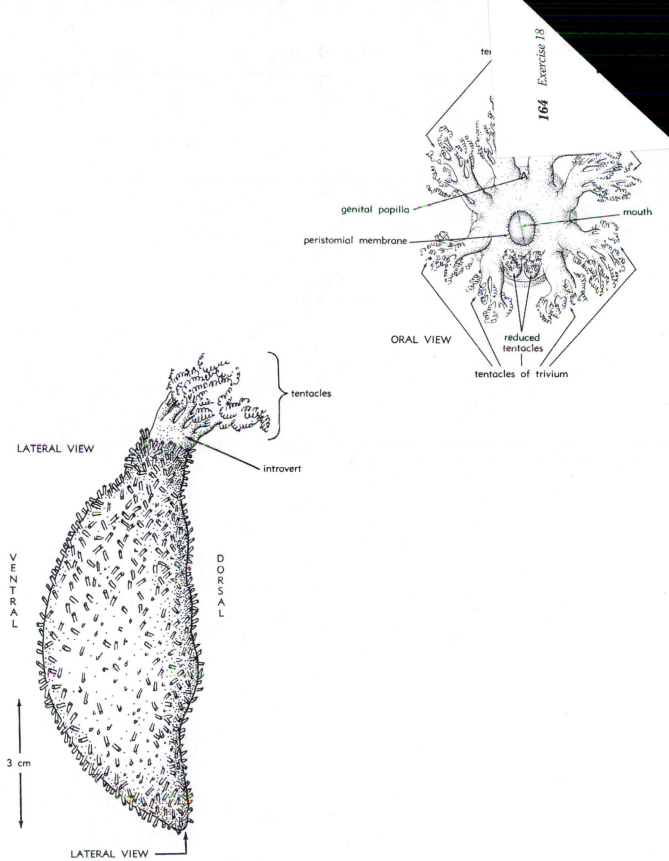

genital papilla

peristomial membrane

mouth

reduced tentacles

tentacles of trivium

ORAL VIEW

tentacles

LATERAL VIEW

introvert

VENTRAL

DORSAL

3 cm

LATERAL VIEW

FIGURE 18.4   Sea Cucumber (External)

Questions

1. Do sea stars have an anterior end?

2. Of what use are the pedicellariae?

3. Are the spines a portion of an endoskeleton or exoskeleton?

4. Explain the function of the pyloric caeca.

5. How are the tube feet able to pull open bivalve molluscs?

6. Explain the physiology of the water-vascular system.

7. Of what economic impact are the echinoderms?

Unit IV
A Survey of Vertebrate Organ Systems

EXERCISE 19 **Phylum Chordata**

Learning Objectives

- ✔ Describe the 4 basic characteristics of the chordate phylum.
- ✔ Identify the primary characteristics which separate invertebrates, chordates and vertebrates.
- ✔ Recognize members of each of the vertebrate classes.

Introduction

The invertebrate animals, previously studied, represent successful members of various lower phyla. Whether simple or complex; soft-bodied or protected by a skeletal structure; sessile or locomotive, they lack the major features of the most highly evolved phylum, the **Chordata**. At some time in their life cycle, members of this phylum possess the following: (1) a stiffening rod-like structure or **notochord**, (2) a **dorsal, hollow nerve cord**, (3) paired, pharyngeal **gill slits**, and (4) a **postanal tail**.

Chordates also retain the familiar features of: bilateral symmetry, complete digestive tract, segmentation, cephalization and deuterostome development. Three subphyla (Urochordata, Cephalochordata and Vertebrata) contain an array of organisms whose embryological development is very similar but whose adult forms vary dramatically. Their organ-systems have evolved as life emerged from aquatic to terrestrial environments. Sometimes referred to as the **protochordates** (proto; "first"), both urochordates and cephalochordates are the nonvertebrate groups. The Craniata or Vertebrata possess one more structural feature: a bony or cartilaginous **skeleton** surrounding the nervous system. Hence, the name derives from the protection around the brain (**cranium**) and the spinal cord (**vertebrae**).

Subphylum Urochordata

The urochordates are unique animals that live in the sea either as solitary organisms or in colonial forms. Metamorphosis transforms a free-swimming, tadpole-like body with the four chordate characteristics into a sac-shaped bag usually attached to the ocean floor. Observe a specimen of an adult sea squirt (*Molgula*). Note the delicate appearance of its covering, the **tunic**, composed of a polysaccharide similar to cellulose. Locate two, tubular **siphons**: the incurrent brings a water stream into a modified pharynx having gill slits while the excurrent forcefully discharges the filtered water. Because of these features, urochordates are also known as **tunicates** or **sea squirts**. The nervous system is reduced to a single elongated neural ganglion near the incurrent siphon. There is no evidence of a notochord or postanal tail in the adult form (Plate 19.1).

Larval sea squirts are free-swimming with the stiffening skeletal rod or **notochord** confined to the tail extension. A dorsal nerve cord runs along the length of the notochord. Although there is no distinctive head region, food and water enter the mouth and pass into the slitted pharynx. Water flow exits a second opening, the **atriopore**. The dramatic transformation from larva to adult results in an equally dramatic loss of most of the chordate characteristics. Lacking a skeletal covering around the nerve cord identifies the urochordates as invertebrates.

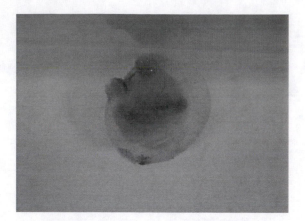

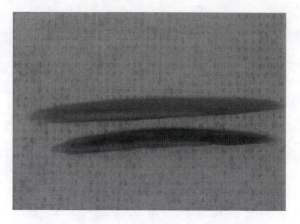

PLATE 19.1   Urochordata (Sea Squirt)

PLATE 19.2   Cephalochordate (Amphioxus)

Subphylum Cephalochordata

Example: *Branchiostoma (Amphioxus)*

The **lancelet**, *Branchiostoma* or *Amphioxus,* represents the cephalochordates; so-called because the notochord extends from the head to the tail region. They represent an evolutionary stage in which chordate characteristics are retained throughout the entire life, but vertebrate characteristics are absent. Living along sandy shores, their transparent bodies filter detritus through prominent gill slits. Obtain a preserved lancelet and scan its length using a hand lens or dissecting scope. Note the anterior end has a fringed covering of **cirri** (sing. cirrus) which contains chemoreceptors and helps filter large particles. Water enters the mouth, which is invaginated into a protective chamber or **oral hood**. It exits through the **atriopore**, an opening about two-thirds down the length of the animal body. Distinctive V-shaped blocks of muscle, the **myotomes**, are located on both sides of the body. Below the myotomes, mature **gonads** are visible through the epidermis. Extended dorsal and ventral **fins** can be traced posteriorly to the **caudal fin** forming the **post-anal tail** (Figure 19.1, Plate 19.2).

In a cross section slide of *Branchiostoma,* locate the dorsal fin ray and progress ventrally to identify the: nerve cord, notochord and chambered pharynx. The lateral pharyngeal walls have supporting **gill bars** separating the open **gill slits**. In a cut made at the pharyngeal level, the **myotome** segments progress in block-like form. Surrounding the pharynx but on the ventral surface, locate the two large **gonad** chambers (Figure 19.2).

Subphylum Vertebrata

The addition of a cartilaginous or bony skeleton around the nervous system clearly separates and identifies the vertebrate animals. Ranging from the elongated bodies of the agnaths to the four-limbed mammals, vertebrates are found in all types of habitats. They have modified their organ-systems for life in the water, on land and even for flight.

Superclass Agnatha

Jawless Fish

Hagfish and **lampreys** have eel-like bodies with a cartilaginous skeleton and a persistent notochord. They lack jaws but their mouths may be surrounded by well developed teeth or suckers. Locate specimens of both on the display table. Marine hagfish burrow in the mud feeding primarily on polychaete worms and dead fish. Their integumentary system has large secretory glands capable of producing copious amounts of slimy **mucus**. Lampreys use **suckers** and cutting **teeth** to attach to fish for a blood meal. Marine lampreys are studied for their migratory routes from salt water to fresh water streams for spawning. Look at the mouth area of preserved agnaths and note their primitive fish form (Plate 19.3).

Superclass Gnathostomata (Jawed Vertebrates)

Class Chondrichthyes

The sharks, skates and rays are included in the chondrichthyes because of their **cartilaginous** skeleton. Paired **fins**, heterocercal tail and a **ventral mouth** help to distinguish these fish from the bony ones. Another external feature to look

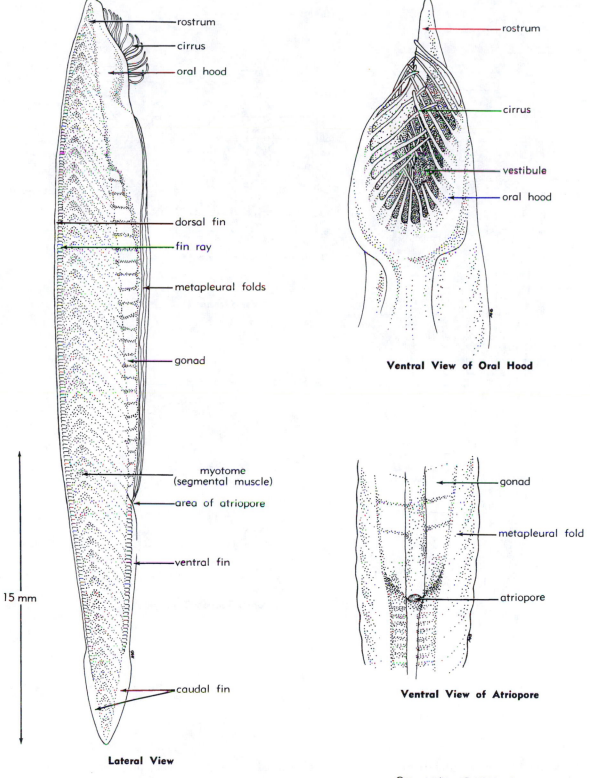

Lateral View

Ventral View of Oral Hood

Ventral View of Atriopore

FIGURE 19.1 Amphioxus (External View)

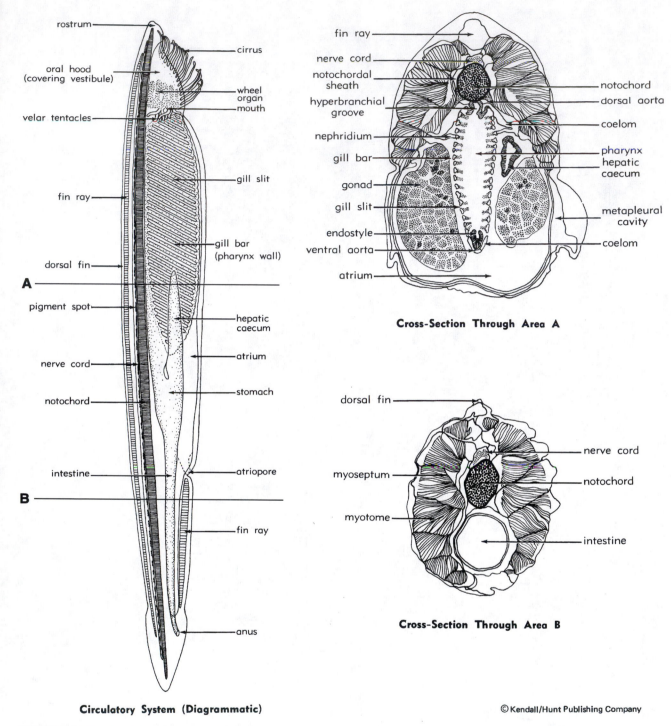

Circulatory System (Diagrammatic)

Cross-Section Through Area A

Cross-Section Through Area B

© Kendall/Hunt Publishing Company

FIGURE 19.2 Amphioxus (Internal View)

for are the slitted **gill** openings behind the mouth which are uncovered. Touch the skin of a preserved shark specimen and note its abrasiveness due to the enamel-like **placoid scales** (Plate 19.4).

Class Osteichthyes

The bony fish are represented by familiar hobby, game and food varieties such as; perch, tuna, guppy, catfish, salmon, trout and eel. Look at the various fish specimens on the demonstration table and point out the paired fins, **homocercal tail**, **terminal mouth** and bony **operculum** covering the gills. If a slide is available, observe the differences between

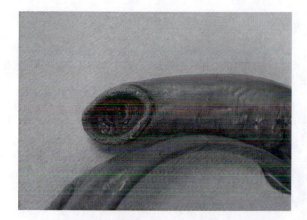

PLATE 19.3   Agnatha (Lamprey)

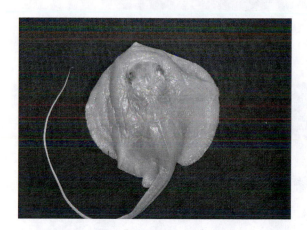

PLATE 19.4   Chondrichthyes (Ray)

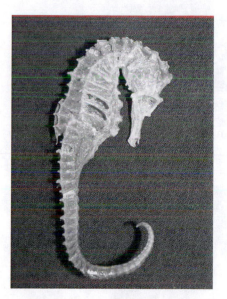

PLATE 19.5   Osteichthyes (Sea Horse)

the types of scales: **ganoid, cycloid** and **ctenoid.** If a fish skeleton has been cleaned and mounted, look at the skeletal system composed of **bone** (Plate 19.5).

Class Amphibia

When adapting to terrestrial existence, amphibians modified their body for: air breathing, prevention of dessication, fluctuating temperatures, and movement on land. They still retain their connection to a watery past because of their mode of reproduction. Eggs are shed in aquatic environments and the young generally pass through a gill breathing, swimming, **tadpole** stage before converting to limbed, air-breathing land dwellers. Most adult amphibians have four limbs (**tetrapods**) although the **caecilians** have a slender worm-like body. Their skin is **moist,** often with poison glands, and their head and trunk are fused with no neck region. **Salamanders** and **newts** are carnivorous amphibians that possess a tail. Observe live or preserved varieties such as *Ambystoma* (axolotl) or *Necturus* (mudpuppy). In *Necturus,* the sexually mature adult retains the larval gill features; a condition called **paedomorphosis** (Plate 19.6).

Frogs and toads have specialized hind limbs for jumping and leaping. They pass through a transformation that absorbs the tail, grows limbs and develops internal lungs. The leopard frog, *Rana pipiens,* is the typical amphibian used in biology laboratories and will be studied thoroughly in the next exercises.

Class Reptilia

The body of reptiles is covered with thickened epidermal **scales.** Members are lung-breathers and may be limbless or tetrapods. They are freed from the aquatic dependence for reproduction because they produce a calcareous or leathery,

PLATE 19.6   Amphibia (Salamander)

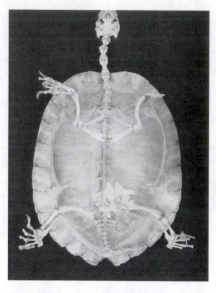

PLATE 19.7   Reptilia (Turtle)

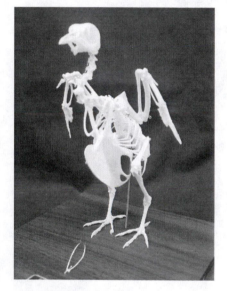

PLATE 19.8   Aves (Bird)

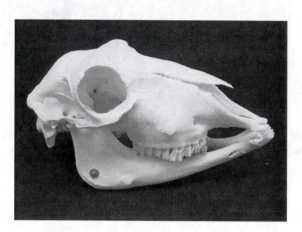

PLATE 19.9   Mammalia (Skull)

shelled egg. Four extraembryonic membranes: the **chorion, amnion, allantois** and **yolk sac,** protect the developing embryo and surround it in a watery environment. Look at the scales of limbless snakes and compare the skin to that of the shelled turtles. Lizards, geckos, crocodiles and alligators all possess the common reptilian features (Plate 19.7).

Class Aves

Bird members are easily identified by their body covering of **feathers.** Although flight is not possible in all birds, they have **wings** as forelimbs and **scaled legs.** Look at several types of feathers. Observe one under a dissecting microscope. Ascertain the habitat and diet of different birds on the display table by comparing their beaks and feet. Look for the **keeled** sternum and long neck on a pigeon skeleton. Compare the size, shape and color of various bird eggs (Plate 19.8).

Class Mammalia

All mammals possess **hair, mammary glands,** movable **eyelids** and fleshy **ears.** The mouth, with its **dentition formula,** is a means of identification in mammalian orders. Check out various skulls and look for differences in their teeth shape and numbers. If taxidermy preparations of skins are available, be able to recognize some common members. Compare the skeleton of a cat to a bird, snake, turtle, frog and fish (Plate 19.9).

Questions

1. Of the vertebrate members, which groups are predominantly aquatic? Predominantly terrestrial?

2. Describe the major problems for multiple systems when changing from life in the water to a terrestrial existence.

3. Which of the chordate members produces an amniotic egg?

4. Name the specific three groups (invertebrate or vertebrate) of animals who have achieved true flight.

5. What is the dentition formula for a human?

6. What type of fertilization, external or internal, is found in reptiles and birds with shelled eggs?

7. What is the economic significance of each of the vertebrate groups?

EXERCISE 20 The Frog: Vertebrate Representative

Learning Objectives

✔ Explain how vertebrate form and function are influenced by each other.
✔ Identify the external parts of the frog body.
✔ Identify the differences in the external covering of amphibians to those of other vertebrate groups.

Phylum Chordata

Subphylum Vertebrata

Example: *Rana pipiens*

External Anatomy

Obtain a preserved frog, a piece of string, a tag, and a dissecting pan. Write the laboratory section and your name on the tag in **pencil** and tie it around the right ankle of your frog. Since the same frog will be used for consecutive days, do not dissect a part of this animal unless instructions are given. After completion of each laboratory period, place the tagged frog in the proper storage jar. Drape the stringed tag over the edge for easy identification. Make sure the frog is completely submerged in the preserving solution to reduce desiccation.

Observe living frogs and concentrate on their breathing and hopping skills. The enlarged **hindlimbs** are powerful and muscular for locomotion. Calculate the average hop distance for a particular specimen. Watch the raising and lowering of the floor of the mouth. The lowering draws air through the nasal openings, the **nares**. As the floor is raised up, the **glottis** opens and air is forcefully pushed into the lungs.

Place the preserved frog in a dissecting pan, belly side down (Figure 20.1). Note the lack of a neck region. The body of a frog is divided into a head and **trunk**. Its skin is distinctly marked with pigmented areas or **chromatophores**. In the genus *Rana*, the pigments form irregular spots, giving members the name **leopard frog**. The pigments can increase or decrease in concentration allowing for color changes. This cryptic coloration camouflages the frog while obscuring its shape against the pond background. At the tip of the face, locate the two external **nares**. The bulged **eyes** have a simple upper **eyelid** but the lower lid has a transparent flap. This **nictitating membrane** wipes the eyeball and protects it when swimming with the eyes open. Use forceps to grab the lower lid and stretch this membrane. Just behind the eye is a large round eardrum, the **tympanic membrane** (Figure 20.2, Plate 20.1).

The short forearms have an upper **arm**, elbow, **forearm**, wrist and **hand**. There are only four fingers on the front limbs. Although they appear to have a thumb, it really is equivalent to our second digit, or pointer finger. During courting season, the thumb pad of mature males becomes enlarged and darkly pigmented. These **nuptial pads**, are used to grasp the female during breeding. Just dorsal and central to the junction of the legs is the **cloacal opening** leading into a common chamber for the digestive, urinary and reproductive deposits.

The hindleg sections include a wide muscled **thigh**, **shank** or lower leg, ankle and **webbed feet**. Note the five elongated toes between the webbing. In some African frog species, the toes are also clawed.

Buccal Cavity (Mouth)

With scissors, make a short cut (about 1/2 in.) through the skin and bones at the angle of the left and right jaw. Open the frog's mouth as far as possible. The mouth or **buccal cavity** is enclosed by both the upper and lower jaw (Figure 20.3). Use a dissecting probe and run it along the edge of the upper jaw. The rasping sound indicates the presence of very small **maxillary teeth**. Two knobby protrusions, the **vomerine** teeth, are used to hold the prey just before swallowing. On either side of these teeth are the **internal nares**. The bottom jaw lacks teeth but has the large **tongue** attached to its floor. The preserved tongue is very contracted and short but when alive it is very flexible and can stretch to surprisingly great lengths. Look for the attachment point. Unlike ours, the tongue is connected at the front of the jaw and has a free, forked end at the back of the throat.

Holding the mouth open as far as possible, clean away any debris using paper towels or the tip of the scissors. Deep at the back of the throat lies the **esophagus**. Press on the floor of the bottom jaw near the esophagus causing the slit of the **glottis** to be exposed. On either side of the esophagus are two **Eustachian tube** openings that connect to the tympanum. If the specimen is a male, two **vocal sacs** are situated just below these openings.

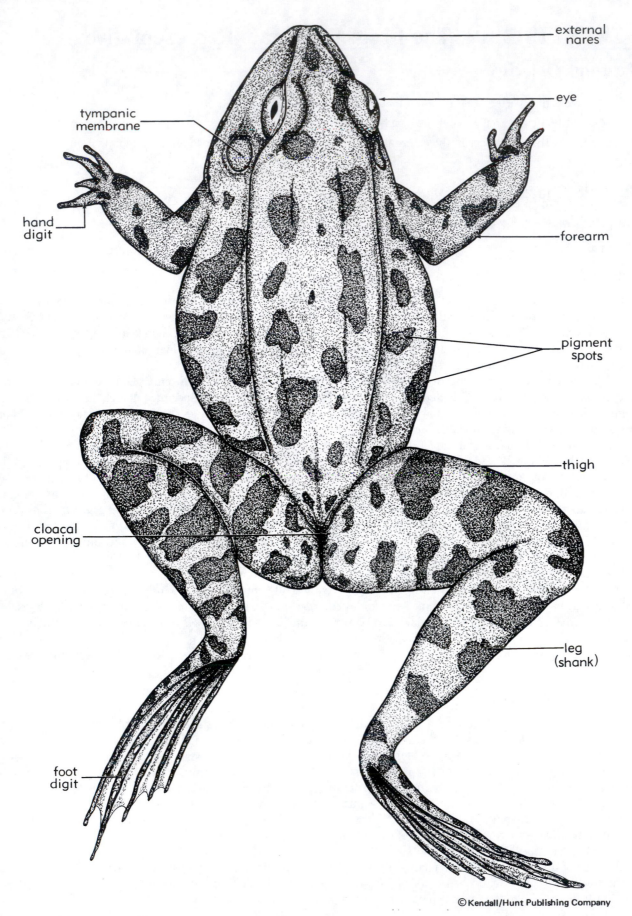

external
nares

eye

tympanic
membrane

hand
digit

forearm

pigment
spots

thigh

cloacal
opening

leg
(shank)

foot
digit

FIGURE 20.1   Frog (External View)

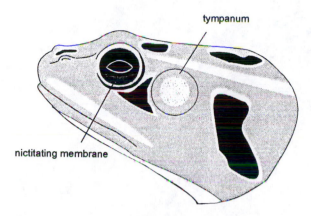

FIGURE 20.2   Lateral view of head

PLATE 20.1   Frog (Head Region)

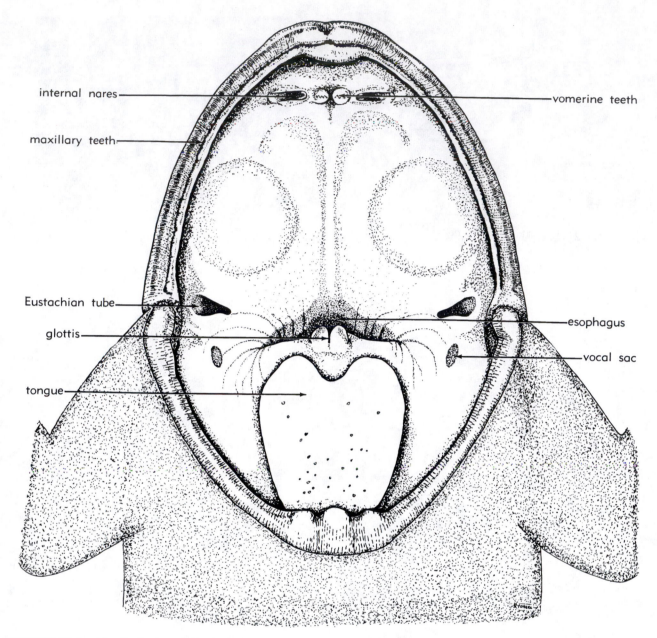

internal nares

vomerine teeth

maxillary teeth

Eustachian tube

esophagus

glottis

vocal sac

tongue

FIGURE 20.3   Mouth Cavity of a Male Frog

Questions

1. What is the benefit of the coloration differences on the dorsal and ventral surfaces?

2. How is the amphibian skin suited to its requirement for a watery habitat?

3. Why do males have an enlarged thumb pad?

4. Why do males have vocal sacs?

5. What is the advantage of the unusual tongue attachment?

6. What purpose does the nictitating membrane serve?

7. What other animals possess a cloaca? Do the mammals?

EXERCISE 21   The Frog: Skeletal System

Learning Objectives

✔ Identify the various bones of the frog skeletal system.

✔ Describe the major changes that occurred in the vertebrate skeleton in adapting to land from an aquatic environment.

✔ Identify individual bones of the human skeleton.

✔ Differentiate between bones of the human axial and appendicular skeleton.

Introduction

Since a major portion of the body is water, aquatic animals are about the same density of water. Therefore, these animals need little physical support compared to that of terrestrial animals. Some aquatic animals, corals and molluscs for example, make use of $CaCO_3$ to provide protection and support. All worms use a hydrostatic skeleton which will suffice in either water or on land since they contact the surface along their entire body. Other land dwelling animals that employ a skeleton include the arthropods and the vertebrates. The arthropods are covered by an external skeleton of chitin and protein which is continuous over their entire body and forms appendages that are flexible. To achieve this flexibility, the skeleton is thinner at the joints than the rest of the skeleton. Vertebrates also have a jointed skeleton. However, the vertebrate skeleton is internal and consists of separate structures (bones) tied together by connective tissues. Because the body is more dense than air and is held off the ground, the vertebrate skeleton must be stiff and provide the necessary support against the effects of gravity. To achieve rigidity, the vertebrate skeleton is composed of $Ca_3(PO_4)_2$ and $CaCO_3$.

Frog Skeleton

Obtain a disarticulated frog skeleton and assemble the bones into two groups. The **axial** and **appendicular** portions separate the body into the skull and midline bones (axial) and the forelimb and hindlimb bones (appendicular) (Figures 21.1 and 21.2).

The most anterior part of the axial skeleton is the **skull** and the most prominent feature is the large **orbit** area in which the eyes sit. The lower jaw, **mandible**, lacks teeth while the **maxilla** bone contains many small teeth. Two knobby teeth of the **vomer** bone are centrally located in the anterior, facial region. The **parasphenoid** is a cross-shaped bone on the ventral surface. A large opening, the **foramen magnum**, opens at the very posterior of the skull. Through this hole, the spinal cord continues from the skull down into the vertebral column.

Behind the skull, a number of **vertebrae** extend. Their numbers vary within the amphibian group and may range from 10 to almost 200. The single **sacral** vertebra attaches to the pelvic girdle. In the frog, a single element, the **urostyle**, extends to the posterior part of the pelvis. Note the absence of ribs in the axial skeleton. The **sternum** lies along the ventral midline and articulates with the pectoral girdle.

The pectoral girdle is composed of the **scapula** bone which extends laterally on the dorsal side and the **clavicle** which extends laterally on the ventral side. Attached at the arm "socket" is the **humerus**. The forearm is a fused **radioulna** bone. The wrist, or **carpal** bones, articulate with the metacarpals and the **phalanges** form the digits.

The pelvic girdle is comprised of three fused sections: the **ilium**, **pubis** and **ischium**. At the hip socket or **acetabulum**, the femur head articulates. A fused limb bone, **tibiofibula**, forms the shank portion of the leg. Two specialized tarsal bones, the **calcaneus** and **astragalus**, form a distinctive curve at the back heel. More **tarsals**, **metatarsals** and **phalanges** complete the lower limbs of the appendicular skeleton.

Human Skeleton

Study Figures 21.3 and 21.4 of the human skeleton and compare it to that of the frog. Then using articulated skeletons or individual bones, identify each of the following and their location in either the axial or appendicular portion (Plates 21.1 A–C, 21.2, 21.3, 21.4 A & B, 21.5, 21.6 A & B).

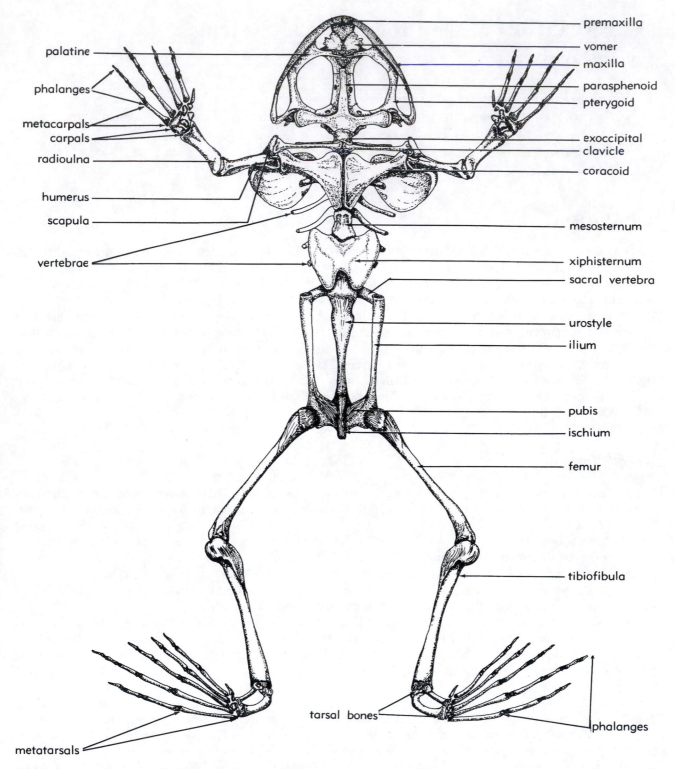

FIGURE 21.1 Frog Skeleton (Ventral View)

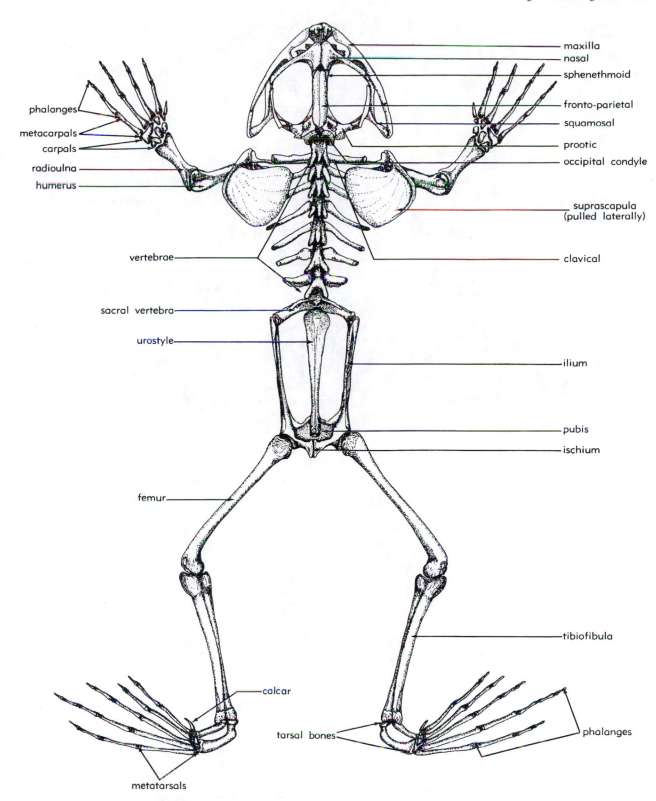

FIGURE 21.2 Frog Skeleton (Dorsal View)

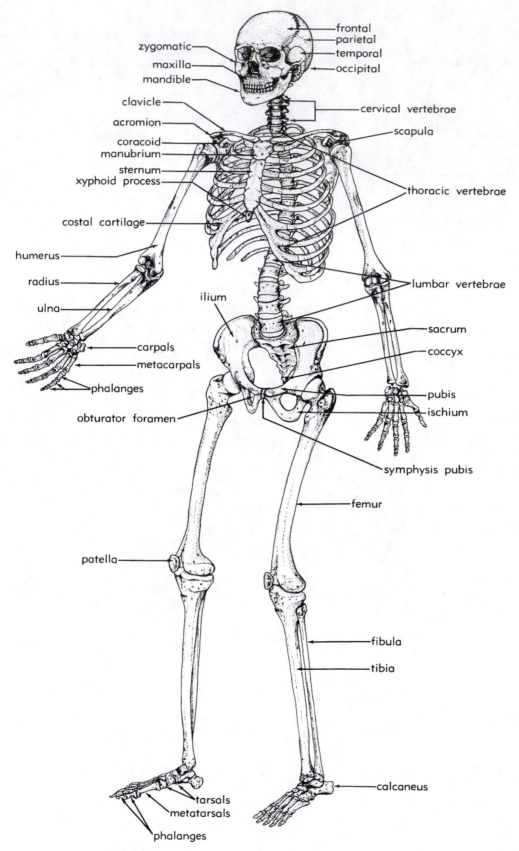

FIGURE 21.3   Human Skeletal System

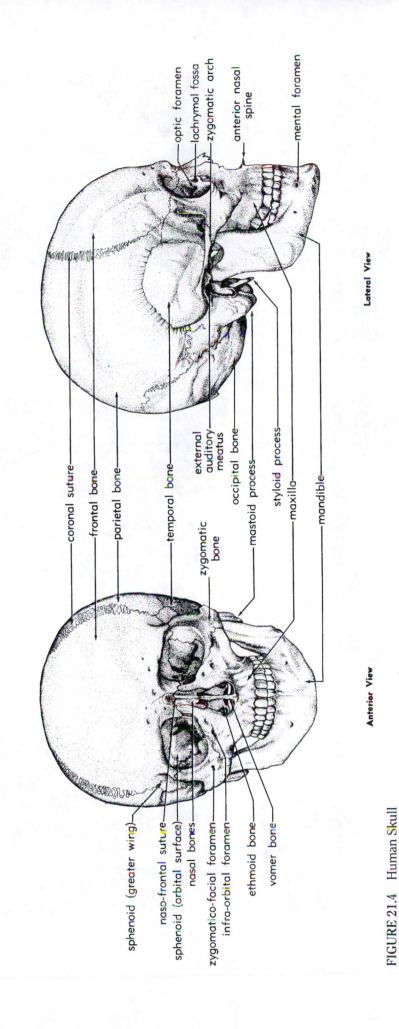

optic foramen
lachrymal fossa
zygomatic arch
anterior nasal spine
mental foramen

Lateral View

coronal suture
frontal bone
parietal bone
temporal bone
zygomatic bone
external auditory meatus
occipital bone
mastoid process
styloid process
maxilla
mandible

Anterior View

sphenoid (greater wing)
naso-frontal suture
sphenoid (orbital surface)
nasal bones
zygomatico-facial foramen
infra-orbital foramen
ethmoid bone
vomer bone

FIGURE 21.4   Human Skull

183

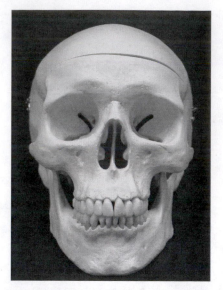

PLATE 21.1 A   Skull (Frontal)

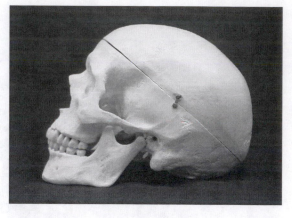

PLATE 21.1 B   Skull (Lateral)

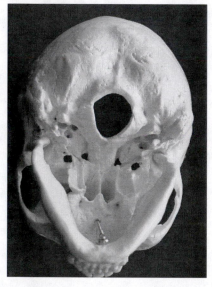

PLATE 21.1 C   Skull (Inferior)

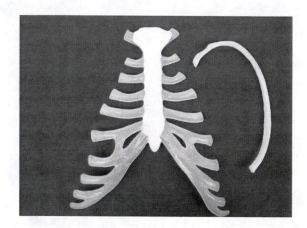

PLATE 21.2   Sternum & Rib

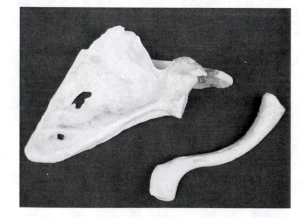

PLATE 21.3   Pectoral Girdle

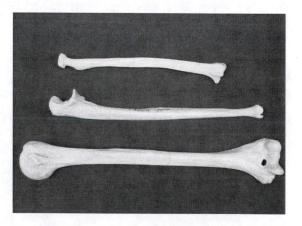

PLATE 21.4 A   Arm Bones

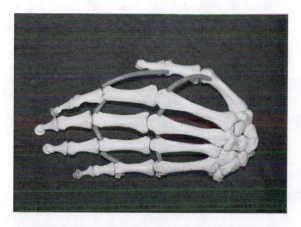

PLATE 21.4 B   Hand Bones

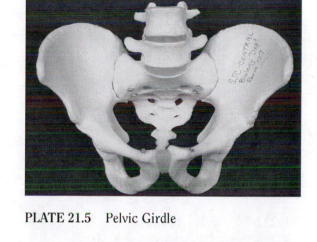

PLATE 21.5   Pelvic Girdle

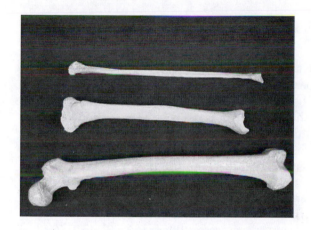

PLATE 21.6 A   Leg Bones

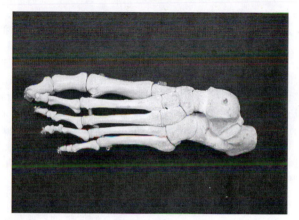

PLATE 21.6 B   Foot Bones

Axial		Appendicular	
Skull	Frontal	Pectoral Girdle	Scapula
	Parietal		Clavicle
	Occipital	Arm	Humerus
	Temporal		Radius
	Zygomatic		Ulna
	Nasal		Carpal
	Ethmoid		Metacarpal
	Vomer		Phalanges
	Maxilla	Pelvic Girdle	Innominate
	Mandible	Leg	Femur
Throat	Hyoid		Patella
Trunk	Sternum		Tibia
	Ribs		Fibula
	Vertebrae		Tarsal
	Sacrum		Metatarsal
	Coccyx		Phalanges

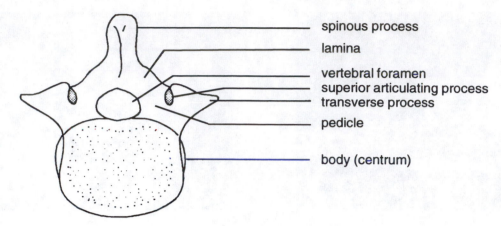

FIGURE 21.5   Typical vertebrae

Additional Axial Bone Features

There are some additional features to identify on the skull which are not whole bones. On the lateral side of the skull at the temporal bone, locate the **external auditory meatus**, or opening of the ear canal. Another larger hole, **foramen magnum**, is located at the base of the skull. The spinal cord exits here and then enters the vertebral column. The **mastoid process** is a large protrusion at the base of temporal bone. The **styloid process** is a pointed projection just below the auditory meatus. The **anterior nasal spine** forms a pointed extension just below the nasal opening.

Obtain an articulated vertebral column or separate vertebrae. Starting at the anterior end, count the different types of vertebrae and numbers of each. The first **cervical** vertebrae (C1), the **atlas**, articulates with the skull. The second cervical vertebrae (C2), the **axis**, has a recognizable knobby extension, the **dens**. The rest of the cervical vertebrae are numbered C3 to C7 and can be identified by their three foramina. The next set of 12 bones, **thoracic** vertebrae (T1–T12) are associated with the ribs. Another set of 5 **lumbar** vertebrae are strong, big-bodied bones.

Using any type of vertebrae, look for the lateral extensions or **transverse** processes (Figure 21.5, Plate 21.7a–d). A single **spinal process** points outwards dorsally. The central hole for the spinal cord is the **vertebral foramen**. The large body or **centrum** gives the vertebrae their bulk. Count the 12 pairs of ribs on an articulated skeleton. The first seven pairs are called **true ribs**, the last five pairs are called "false ribs" of which the last two pairs are also called "floating ribs".

Looking at the sternum, distinguish between its three parts: the upper **manubrium**, the **body** of the sternum and the **xiphoid process**. Distinguish three sections of the pelvic bone as well. The **ilium**, **ischium** and **pubic** portions form a saddle-like pelvic girdle when attached to the sacral vertebrae. Identify the **obturator foramen** and the socket or **acetabulum** (Figure 21.6).

Additional bones of the skull which can not be seen because they are embedded within the temporal bone are the three ear **bones**. Locate a box of these small bones and note their shapes give them logical, common names. The **malleus** or hammer, appears to have a handle and rounded head portion. The **incus** looks like a blacksmith's anvil, and the **stapes** is shaped like a stirrup.

Additional Appendicular Bone Features

Obtain a femur bone from the demonstration table that has been split longitudinally to expose the inside. The ends of the bone, **epiphyses**, are separated by a long shaft, the **diaphysis**. The hollowed center of the bone would normally be filled with **bone marrow**. Two types of bone can also be seen. The sides of the shaft are solid, thick, **compact** bone to give support, while the heads are airy and composed of **spongy** bone.

PLATE 21.7 A   Cervical (Atlas)

PLATE 21.7 B   Cervical (Axis)

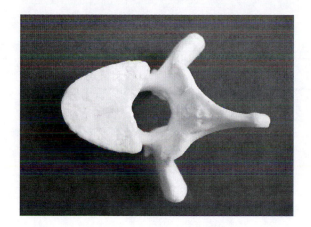

PLATE 21.7 C   Thoracic

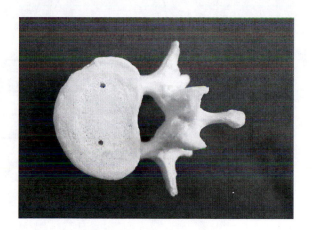

PLATE 21.7 D   Lumbar

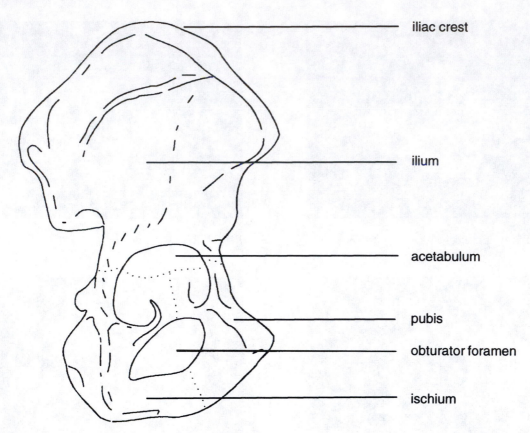

iliac crest

ilium

acetabulum

pubis

obturator foramen

ischium

FIGURE 21.6 Pelvic (innominate) Bone

Questions

1. List four functions of the vertebrate skeleton.

2. Why do aquatic vertebrates have more vertebrae than terrestrial vertebrates?

3. Why are the vertebrae of the human skeleton progressively larger from top to bottom?

4. In what ways do female and male human skeletons differ?

5. Which skeletal system gives greater protection, frog or man? Why?

6. In which of the two portions; axial or appendicular, are the most human bones found?

EXERCISE 22   The Frog: Muscular System

Learning Objectives

- ✔ Describe the general functions of all muscles.
- ✔ Explain how muscles provide strength and continuous movement during contraction.
- ✔ Identify major muscles of the frog.
- ✔ Identify major muscles or muscle groups of the human.

Introduction

The muscles make up the bulk of the amphibian and human body. When contracting, they: make movement possible, assist in breathing, pump the blood and move food through the digestive tract. As previously studied in Chapter 2, there are three categories of muscle tissue: smooth, cardiac and skeletal. To study physical movement, the skeletal muscles attached to bone must be identified, named and assigned a specific action. Since it is not in the realm of this class to study every muscle, only a select few will be dissected.

Dissection Procedure

Obtain your frog from the preserving canister and place it in the dissecting pan with the ventral side up. Make a small scissor cut at the **midline** of the ventral (belly) side of the body. From that point, cut along the midline down to the cloacal opening and up to the tip of the lower jaw. From this original incision, cut through the skin from the midline in the chest region down each forelimb to the wrist and from the midline in the abdominal region down each hind limb to the ankle.

Remove the skin by pulling it from the underlying muscles and joints. The skin at the joints is tightly bound and will have to be snipped away with scissors. When removing the skin, be careful not to remove muscle tissues, especially the small muscles of the lower thigh area. The skin cuts on the arms and legs may become loose enough to be able to grab the skin and "peel" it off.

Along the dorsal (back) side, the skin is more closely attached. Again separate the skin away by making a mid-dorsal cut from cloaca to just behind the eyes. Similar side cuts down the arms and legs should allow large sections of skin to be removed easily. On the head, use forceps to pick up the skin and snip it away. Make a circular cut around each eye and leave the skin covering around each eyeball and the tympanum.

Before discarding the skin pieces in the medical waste barrel, look at the underside and note the large number of blood vessels within it. This rich vascular supply is necessary for the gas exchange that occurs across the skin.

Identification of Muscles

Look at the diagram of the frog's muscle in Figure 22.1 and 22.2. Note the location (dorsal/ventral, head, trunk, leg, etc.) and general shape of the superficial muscles described below. Dissecting will require that they are separated from each other by teasing away the tissue covering or **fascia**. Isolate but do **NOT** cut any muscles. Try and find the connection points of both ends. Skeletal muscles have an **origin**, or more fixed point attached to one bone. The **insertion** or more movable end is attached to another bone across a joint. The **belly**, or middle of the muscle, generally will have a recognizable shape and direction to its fibers (Plate 22.1 A & B).

Head Region

Temporal (Temporalis):

The temporalis is located between the eye and tympanum. It extends dorsally through these organs and helps to close the mouth. If the skin was not properly removed in this area, finish snipping it away to expose this muscle.

Superficial   **Deep**

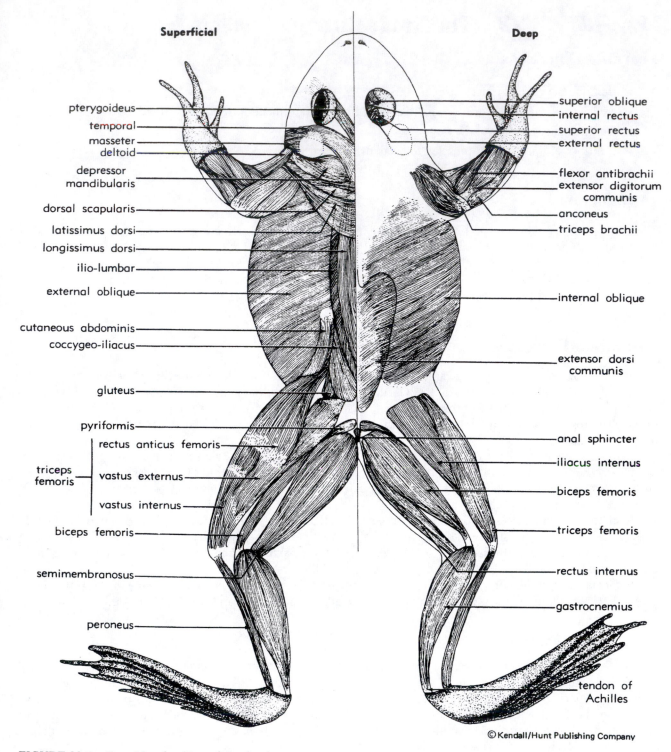

pterygoideus
temporal
masseter
deltoid
depressor
mandibularis
dorsal scapularis
latissimus dorsi
longissimus dorsi
ilio-lumbar
external oblique
cutaneous abdominis
coccygeo-iliacus
gluteus
pyriformis
rectus anticus femoris
triceps femoris
vastus externus
vastus internus
biceps femoris
semimembranosus
peroneus

superior oblique
internal rectus
superior rectus
external rectus
flexor antibrachii
extensor digitorum communis
anconeus
triceps brachii
internal oblique
extensor dorsi communis
anal sphincter
iliacus internus
biceps femoris
triceps femoris
rectus internus
gastrocnemius
tendon of Achilles

© Kendall/Hunt Publishing Company

FIGURE 22.1   Frog Muscles (Dorsal Surface)

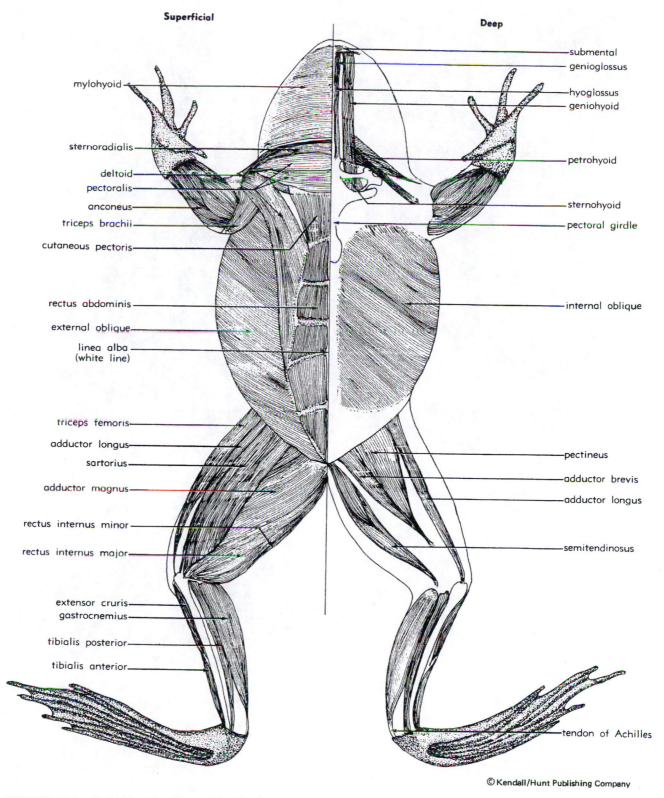

Superficial   Deep

- mylohyoid
- sternoradialis
- deltoid
- pectoralis
- anconeus
- triceps brachii
- cutaneous pectoris
- rectus abdominis
- external oblique
- linea alba (white line)
- triceps femoris
- adductor longus
- sartorius
- adductor magnus
- rectus internus minor
- rectus internus major
- extensor cruris
- gastrocnemius
- tibialis posterior
- tibialis anterior

- submental
- genioglossus
- hyoglossus
- geniohyoid
- petrohyoid
- sternohyoid
- pectoral girdle
- internal oblique
- pectineus
- adductor brevis
- adductor longus
- semitendinosus
- tendon of Achilles

FIGURE 22.2 Frog Muscles (Ventral Surface)

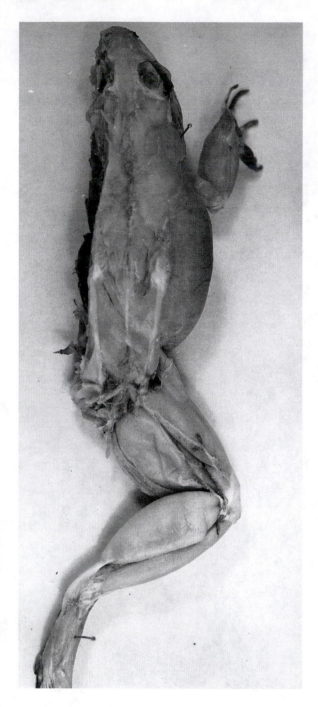

PLATE 22.1 A   Frog Muscles (Dorsal View)

PLATE 22.1 B   Frog Muscles (Ventral View)

Depressor Mandibularis (Depressor Mandibuli):

This muscle is located posterior to the tympanum and extends dorsally. Contraction of this muscle opens the mouth by depressing the lower jaw.

Mylohoid (Mylohyoideus):

The mylohyoid runs transversely across the lower jaw. Its action is important in swallowing and elevation of the floor of the mouth during respiration.

Pectoral Girdle and Forearm Region

Deltoid (Deltoideus):

The deltoideus extends from the anterior border of the upper arm toward the floor of the lower jaw and its action draws the arm forward.

Pectoralis:

The pectoralis is a large fan-shaped muscle located on the ventral surface and extends from the humerus across the chest toward the sternum, coracoid, and rectus abdominus muscle. Its motion activity is arm flexion.

Triceps Brachii:

This muscle is found on the dorsal and posterior side of the upper arm and its action extends the forearm.

Trunk Region

Dorsal Scapularis (Dorsalis Scapulae):

This muscle is located beneath and posterior to the depressor mandibularis and extends dorsally from the upper region of the arm toward the back. It raises the arm toward the body.

Latissimus Dorsi:

The latissimus dorsi extends from the dorsal part of the trunk, or scapula, and attaches to the humerus. It lies posterior to the dorsal scapularis and raises the arm upward and backward (abduction).

Longissimus Dorsi:

The powerful longissimus dorsi muscle runs lengthwise parallel to the vertebrae. This muscle raises the head and straightens the back.

Rectus Abdominis:

The rectus abdominis is a thin sheet of muscle running ventrally from the sternum to the pubis parallel to the linea alba. It appears in sectioned compartments and supports the abdomen.

External Oblique (Obliquus Externus):

This muscle is the lateral portion of the abdomen running posteriorly and ventrally. It can be seen stretched across the bulging sides of the abdomen which it supports and compresses.

Thigh Region

Triceps Femoris:

This grouping consists of three muscles covering the anterior portion of the thigh both dorsally and ventrally. It draws the thigh toward the trunk and straightens the shank in the jumping action.

Semimembranosus:

The large semimembranosus is located on the dorsal surface of the posterior border of the thigh. Its action is to bend the shank.

Rectus Internus Minor (Gracilis Minor):

This muscle is a thin band located on the posterior border of the thigh both dorsally and ventrally. It flexes the shank and adducts the thigh.

Rectus Internus Major (Gracilis Major):

This muscle is posterior to the adductor magnus and partially hidden by the rectus internus minor. It also bends the shank and draws the thigh.

Sartorius:

The sartorius is the most superficial muscle on the thigh. It is a flat, thin band covering the middle portion of the thigh on the ventral surface. It lies over the junction of the triceps femoris and the adductor magnus. It bends the shank and draws the thigh toward the trunk.

Adductor Magnus:

This muscle runs under the sartorius muscle and lies between the rectus internus major muscle and the triceps femoris. As the name implies, it adducts the thigh and the leg.

Biceps Femoris:

The biceps femoris is a small muscle located on the dorsal surface of the thigh between the triceps femoris and semimembranosus. Tease between these two muscles to find the spindle-shaped biceps that flexes the shank.

Shank Region

Gastrocnemius:

The large "calf" muscle located on the posterior portion of the lower leg. Its action extends the foot and flexes the shank. The familiar **tendon of Achilles** is the large white, tendon connection of the gastrocnemius to the floor of the foot.

Peroneus:

The peroneus is a short, stout muscle lying anterior to the gastrocnemius on the dorsal portion of the shank. It can be recognized by the strong tendon that attaches on the femur, crosses the ankle joint and attaches to the tarsal bones. It extends the shank and flexes the foot.

Tibialis anterior (Tibialis anticus):

The **tibialis anterior** lies to the front of the tibiofibula bone. It is a thin muscle that attaches under the peroneus at the knee joint. The two muscles (peroneus and tibialis anterior) may appear as one. Close examination will show the separation line between them. The tibialis anterior also extends the shank.

Tibialis posterior (Tibialis posticus):

The tibialis posterior is attached behind and along most of the tibiofibula bone and is covered by the large gastrocnemius. Tension on the tibialis posterior extends the foot.

Human Muscles

Using the diagrams in Figure 22.3 and 22.4, find and name the muscles in the appropriate areas of your body. Many will be recognizable because of the similarity in location to the frog musculature. Common muscles familiar to athletes and fitness enthusiasts are included in the short list below:

Ventral Surface	Dorsal Surface
Frontalis	Trapezius
Temporalis	Triceps brachii
Deltoid	Latissimus dorsi
Pectoralis major	Gluteus maximus
Biceps brachii	Biceps femoris
Rectus abdominis	Semimembranosus
External Oblique	Gastrocnemius
Sartorius	Tendon of Achilles
Rectus femoris	
Tibialis anterior	

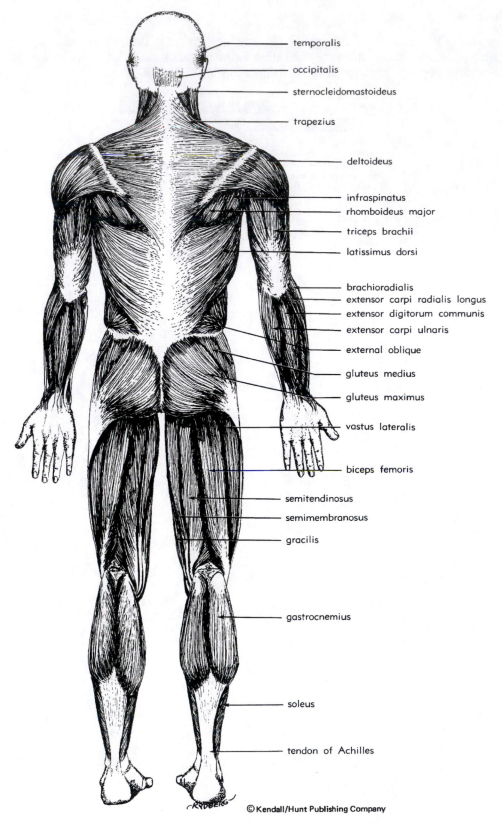

temporalis
occipitalis
sternocleidomastoideus
trapezius
deltoideus
infraspinatus
rhomboideus major
triceps brachii
latissimus dorsi
brachioradialis
extensor carpi radialis longus
extensor digitorum communis
extensor carpi ulnaris
external oblique
gluteus medius
gluteus maximus
vastus lateralis
biceps femoris
semitendinosus
semimembranosus
gracilis
gastrocnemius
soleus
tendon of Achilles

FIGURE 22.3   Human Muscles (Posterior View)

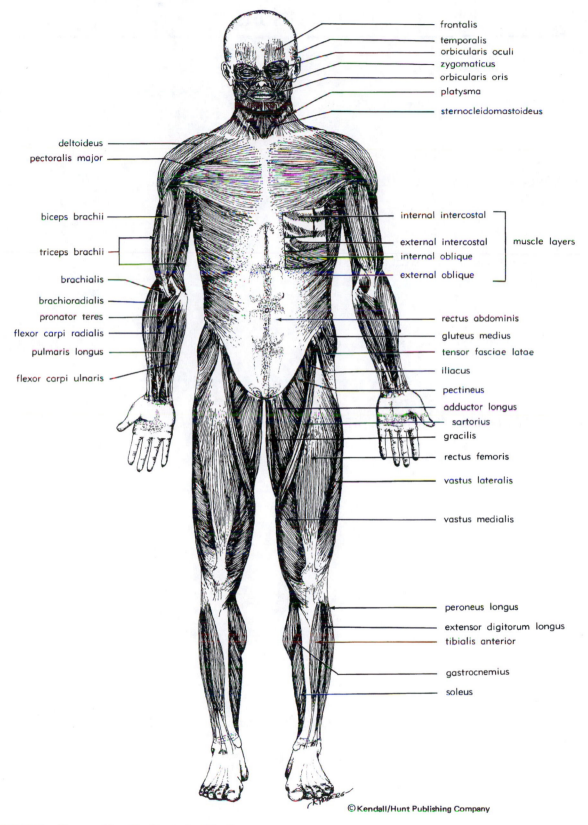

frontalis
temporalis
orbicularis oculi
zygomaticus
orbicularis oris
platysma
sternocleidomastoideus

deltoideus
pectoralis major

biceps brachii

triceps brachii

brachialis
brachioradialis
pronator teres
flexor carpi radialis
pulmaris longus
flexor carpi ulnaris

internal intercostal
external intercostal
internal oblique
external oblique

muscle layers

rectus abdominis
gluteus medius
tensor fasciae latae
iliacus
pectineus
adductor longus
sartorius
gracilis
rectus femoris
vastus lateralis
vastus medialis

peroneus longus
extensor digitorum longus
tibialis anterior
gastrocnemius
soleus

© Kendall/Hunt Publishing Company

FIGURE 22.4   Human Muscles (Anterior View)

Questions

1. Why are muscles arranged in antagonistic pairs?

2. Give some names of antagonistic pairs of muscles.

3. Locate the tendon of Achilles. What is its action?

4. What connects a muscle to the surface of a bone?

5. Why would the leg muscles of a frog and a human differ?

EXERCISE 23   The Frog: Digestive System

Learning Objectives

✔ Distinguish between autotrophic and heterotrophic nutrition.

✔ Identify the various components of the frog digestive system.

✔ Describe the function(s) of the various components.

✔ Identify the various components of the human digestive system.

✔ Explain the role of liver, gall bladder and pancreas in digestion.

Introduction

One of the fundamental properties of living organisms is the possession of a high degree of organization. This organization is brought about by synthesizing large, complex **organic** compounds that are fitted together in very specific arrangements. These include familiar chemicals such as: carbohydrates, lipids, proteins and nucleic acids. Some organisms, those with **chlorophyll** (plants), can synthesize the large organic compounds they need from simple inorganic molecules (CO_2 and H_2O) in the reaction of **photosynthesis**. Since animals lack chlorophyll, they must acquire the necessary organic compounds from outside sources in some form of feeding process.

Dividing organisms by their nutritional source results in two general types; **autotrophs** and **heterotrophs**. Autotrophs, such as the plants who synthesize organic materials in photosynthesis produce their own food. Heterotrophs, like animals, take in an outside source of food material. Because the food matter (meat, fruit, insects, etc.)is too large to be absorbed across cell membranes, food is chemically and physically reduced in size by **digestion**. In the higher organisms, this processing occurs in different organs along the digestive tract.

Digestive System Dissection

Obtain your frog from the storage jar and place it in the dissecting pan, ventral side up. With scissors make a cut in the trunk muscles parallel to and about 1/8 inch from the mid-ventral line. Extend the incision from the crotch to the lower jaw. Making a parallel cut will prevent you from cutting the ventral abdominal vein that lies just below the mid-ventral line. Avoid damaging the underlying organs by keeping the point of the scissors next to the body wall as you cut.

In the chest region, the bones of the sternum will have to be snipped. Continue the midline cut through the mylohyoid muscle to the tip of the bottom jaw. Next, make lateral cuts from the midline to just under the left and right armpits. Repeat another set of lateral cuts from the midline across the groin to the left and right hip line. This forms a "window flap" of body wall which can be folded away and pinned.

Digestive Organs

The internal mouth of the frog was examined in an earlier dissection. Review the mouth area and locate the esophageal opening. Fold back the mylohyoid muscle and tease through the tissue to find the muscular, tube-like **esophagus**. Follow it down under the sternum area.(Be careful not to tear or damage blood vessels in the chest area). Between the lobes of the **liver**, locate a balloon-like bag or **gall bladder**. Lifting the liver lobes should expose the "J" shaped **stomach**. Cut open the stomach. (If swollen and enlarged, food material will be present and insect bodies may be recognizable). Flush the inside wall of the stomach with water and note the folding pattern or **rugae** (Figure 23.1).

At the end of the stomach (away from the mouth), find the constriction known as the **pylorus**. This valve regulates the flow of food into the intestinal tract (Plate 23.1). Continue tracing along the digestive tube and locate the **small intestine**. This is held in position by a membranous **mesentery**. The **large intestine** widens and opens into the **cloaca**. Lift up and examine the mesentery on the inner curvature of the stomach. The **pancreas**, a strip of tissue approximately 1/8 inch wide is embedded near the pyloric valve.

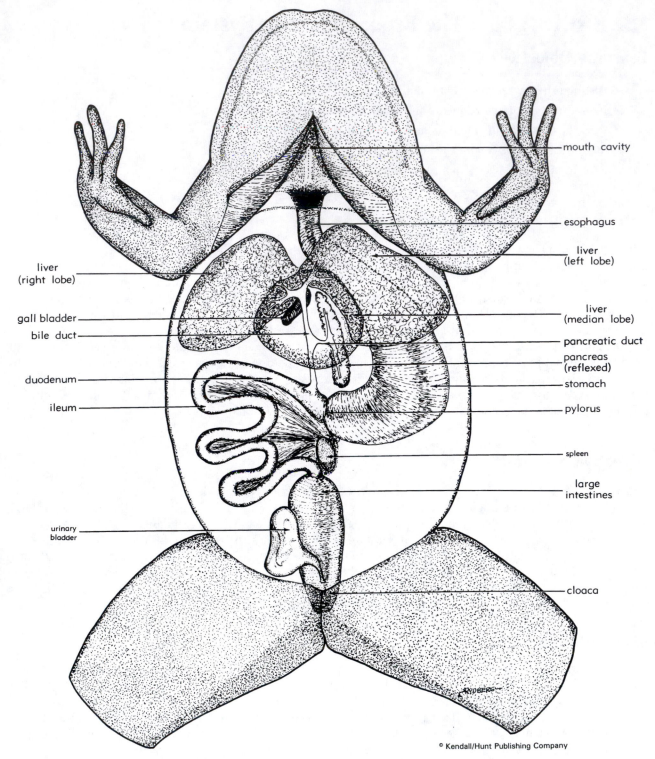

mouth cavity

esophagus

liver
(left lobe)

liver
(right lobe)

liver
(median lobe)

gall bladder

pancreatic duct

bile duct

pancreas
(reflexed)

duodenum

stomach

ileum

pylorus

spleen

large
intestines

urinary
bladder

cloaca

RYDBERG

© Kendall/Hunt Publishing Company

FIGURE 23.1 Frog (Digestive System)

PLATE 23.1   Frog (Internal Organs)

Human Digestive System

Using a human torso or model of the human digestive system, identify the following (Figure 23.2) and their function:

Gastrointestinal Tract	Accessory Organs
Mouth	Teeth
Esophagus	Tongue
Stomach	Salivary Glands
Small Intestine	Liver
Appendix	Gall Bladder
Large Intestine	Pancreas
Rectum	Gastrocnemius
Anus	

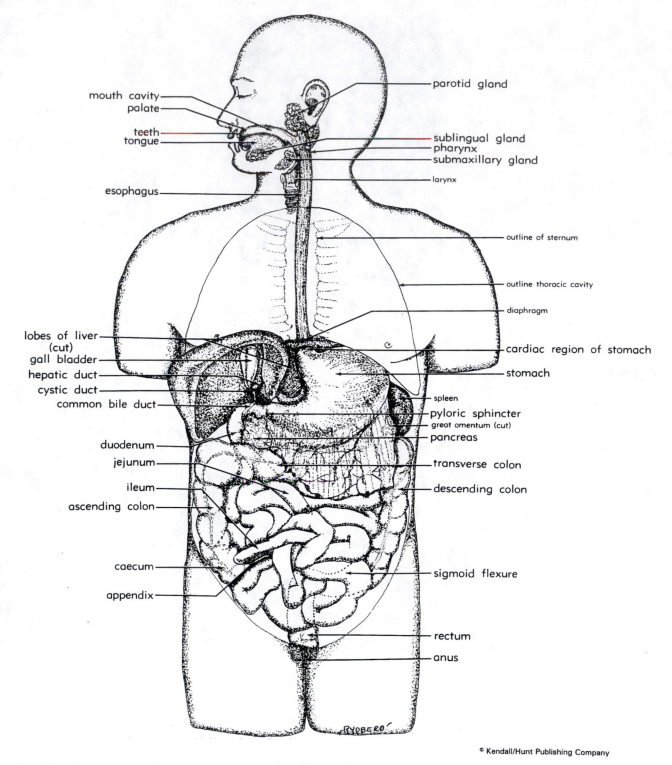

parotid gland

mouth cavity
palate

teeth
tongue

sublingual gland
pharynx
submaxillary gland

larynx

esophagus

outline of sternum

outline thoracic cavity

diaphragm

lobes of liver
(cut)
gall bladder
hepatic duct
cystic duct
common bile duct

cardiac region of stomach

stomach

spleen
pyloric sphincter
great omentum (cut)
pancreas

duodenum
jejunum
ileum
ascending colon

transverse colon

descending colon

caecum

sigmoid flexure

appendix

rectum

anus

© Kendall/Hunt Publishing Company

FIGURE 23.2   Human (Digestive System)

Questions

1. What differences are there in the frog and human mouth?

2. Where do products of the liver, gall bladder and pancreas enter the digestive tract?

3. What is the function of the:
   Liver?

   Bile?

   Pancreas?

4. Explain the difference between a cloacal opening and an anus.

5. What is the function of the:
   Stomach?

   Small intestine?

   Large intestine?

6. What is the role of the pyloric sphincter?

7. How do autotrophs differ from heterotrophs?

8. Identify the subcategories of heterotrophs defined by the type of food ingested.

EXERCISE 24   Digestive Enzymes

Learning Objectives

✔ Explain how mechanical and chemical digestion prepares food for absorption by the body.
✔ Identify what enzymes are involved in chemical digestion and what organ produces these enzymes.

Introduction

Foods that are ingested by most animals must be digested, absorbed, and utilized much in the same manner. A specialized group of proteins, the enzymes, are involved in the chemical breakdown of these food items. Produced by various organs, they target specific groups of chemical compounds and systematically reduce them in size and complexity. The following short experiments are designed to give the general concept of their activities.

Starch Digestion

Obtain a rack with test tubes, paraffin, wax pencil, beaker, graduated cylinder, and test tube holder from the demonstration table. Begin chewing the paraffin at once. (For this experiment two groups may work together.) Collect 4 ml. of saliva in the beaker and add 4 ml. of distilled water to make a saliva solution. **Place 4 ml. of boiled starch into four test tubes and label A, B, C, and D.** Add the following solutions to these tubes:

A. 4 ml. of distilled water.
B. 4 ml. of saliva solution. Stir vigorously.
C. 2 ml. of 0.2% HCl and 2 ml. of saliva solution. Stir vigorously.
D. 2 ml. of 0.5% $NaHCO_3$ and 2 ml. of saliva solution. Stir vigorously.

Determine the pH of each tube by testing with litmus paper and record the pH for each in the chart (Table 24.1). Set the tubes in an incubator at a temperature of 98 F°. After 30 minutes, pipette out one half of the contents of each labeled tube and add to a second clean test tube (also labeled A, B, C, D). Test for starch by adding a few drops of iodine solution to each. A blue-black color denotes starch. Mark the results in the table. Test the remaining half for the presence of simple sugar by adding 10 ml. of Benedict's solution. Heat over a bunsen burner or place in a hot water bath. A green-yellow-red color denotes sugar. Tabulate the results in Table 24.1 (Mark a + for positive and − for negative results).

TABLE 24.1   Starch Digestion

TEST TUBE	pH	STARCH	SUGAR	CONCLUSION
A				
B				
C				
D				

Protein Digestion

The manner in which gastric digestion of protein is broken down into simpler substances will be demonstrated in the following manner. Label five test tubes A, B, C, D, and E. **Place a small piece of boiled egg white (approximately the same size) in each test tube.** Add the following solutions to the test tubes:

A. 20 ml. of 0.1% pepsin solution.

B. 18 ml. of 0.1% pepsin solution and 2 ml. of 0.5% $NaHCO_3$.

C. 18 ml. of 0.1% pepsin solution and 2 ml. of 0.2% HCl.

D. 20 ml. of 0.5% $NaHCO_3$.

E. 20 ml. of 0.2% HCl.

Test the pH of each test tube using litmus paper and record in Table 24.2. Set the tubes in an incubator set at 98 F°. After 50 minutes, observe the appearance of the egg piece in each test tube (If lab time does not allow the completion of this portion, view the test tubes at the next laboratory period). Tabulate the results in the table below:

TABLE 24.2   Protein Digestion

Test Tube	pH	Result	Conclusion
A			
B			
C			
D			
E			

Questions

1. Why are enzymes important to digestion?

2. What is the simple test for starch presence? Sugar presence?

3. What is the name of the enzyme found in saliva?

4. Why is pH important in digestion?

5. What organ(s) are responsible for mechanical digestion?

6. How does mechanical digestion aid chemical digestion?

7. What was the purpose of Test Tube A in the starch set-up?

EXERCISE 25   The Frog: Urogenital System

Learning Objectives

✔ Identify the macroscopic structures of the vertebrate kidney and their functions.

✔ Explain the structure and function of the nephron and how it forms urine.

✔ Identify the function(s) of the various structures of the male and female reproductive system.

Introduction

Reproduction is necessary for the continuation of the species. This depends on the production of sex cells by both sexes. The organs that make sex cells are called the **primary sex organs**. However, there must be a means by which sperm cells can reach and fertilize egg cells in order to produce a new individual. This requires accessory sex organs (secondary) to make fertilization possible.

Due to the various metabolic processes occurring in the body, different kinds of waste products are formed and must be excreted. The function of the kidney is to reduce excessive amounts of these materials to a nontoxic level. Abnormal amounts of water, whether metabolic water or ingested, must be removed. When amino acids are used for energy production or converted to storage compounds such as sugars or fats, the nitrogen group is removed in the form of ammonia. The liver converts ammonia to urea in amphibians and most mammals whereas birds and reptiles convert it to uric acid. In either case, it must be eliminated from the body by the **kidney**.

Frog Urogenital System

In amphibians there is a very close relationship between the excretory and reproductive systems. This involves the use, by both systems, of certain tubes leading to the outside. In all vertebrates, except mammals, a **cloaca** is present at the posterior end of the body. This chamber serves as a common duct to the outside for products of the excretory, reproductive, and digestive systems. Male mammals still use the duct draining the urinary bladder as a common tube for both the excretory and reproductive systems. Because of the close relationship between these two systems, they will be studied together.

Obtain your frog and place in a dissecting pan. The excretory structures are alike in males and females (Figure 25.1 and 25.2). Gently move the small intestine to one side and pin. Observe the two brownish, elongated structures lying on the dorsal surface next to the body wall. The **kidneys** are separated from the coelom by a thin and nearly transparent **peritoneum**. Along the lateral margin of each kidney there is a small duct leading to the cloaca. These are the mesonephric ducts, commonly called **ureters**. They transport the urine from the kidney to the **urinary bladder** in the pelvic area. Urine then drains from the urinary bladder out the cloaca opening. When empty, the urinary bladder resembles a deflated sac.

Within the abdominal cavity of the male frog, **fat bodies** are attached to the kidneys and extend into the coelom on both sides of the intestines (Figure 25.1). The fat bodies store fat and vary in size; usually being larger in the fall and smaller in the spring. Attached to each fat body is a light-colored, bean-shaped **testis**, which produces sperm. The sperm pass from the testis to the kidneys through sperm ducts or **vas efferentia**. From the kidney they continue within **urogenital ducts** and are discharged through the cloacal opening to the outside. The rudimentary (vestigial) oviducts are nonfunctional in males.

The female frog has two white, lobed **ovaries** attached by mesenteries to the ventral side of the kidneys (Figure 25.2). In winter and early spring, the ovaries form distended membranes containing hundreds of black and white **ova** or eggs. In the summer and early fall, the ovaries are small, fan shaped, and light colored. The eggs are released from the ovaries into the coelom. From the coelom they enter two highly coiled **oviducts** through a funnel-shaped structure near the lungs called the **ostium**. The oviducts connect posteriorly to paired **uteri** (ovisac). From each uterus, eggs pass into the **cloaca** and are laid externally. During mating, the male clasps the female on the dorsal side and sheds sperm on top of the eggs as they are released by the female into the water. This embrace is known as **amplexus**.

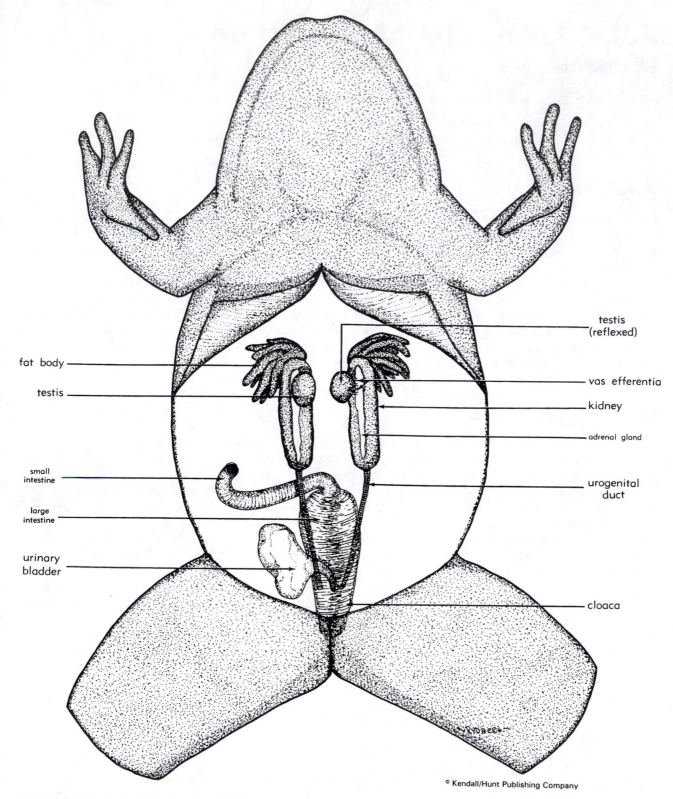

testis
(reflexed)

fat body

testis

vas efferentia

kidney

adrenal gland

small
intestine

urogenital
duct

large
intestine

urinary
bladder

cloaca

© Kendall/Hunt Publishing Company

FIGURE 25.1   Frog (Male Urogenital System)

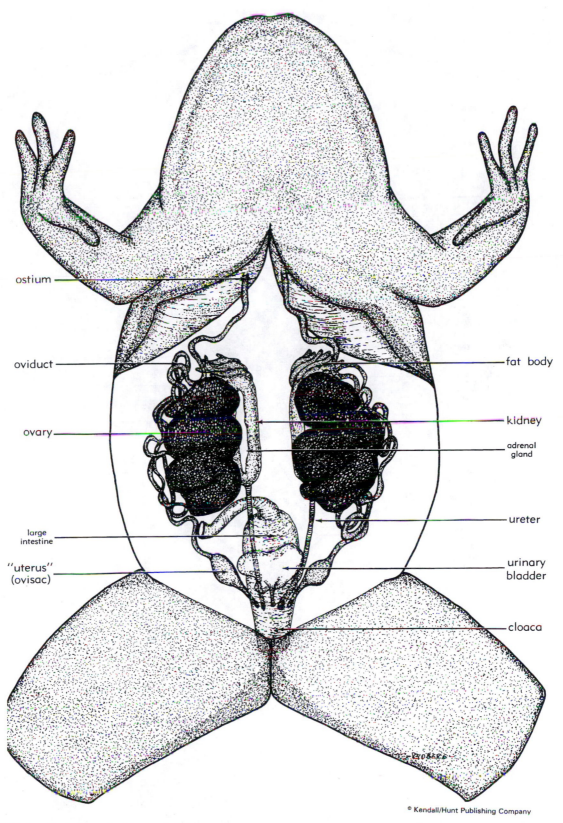

ostium

oviduct

ovary

large
intestine

"uterus"
(ovisac)

fat body

kidney

adrenal
gland

ureter

urinary
bladder

cloaca

FIGURE 25.2   Frog (Female Urogenital System)

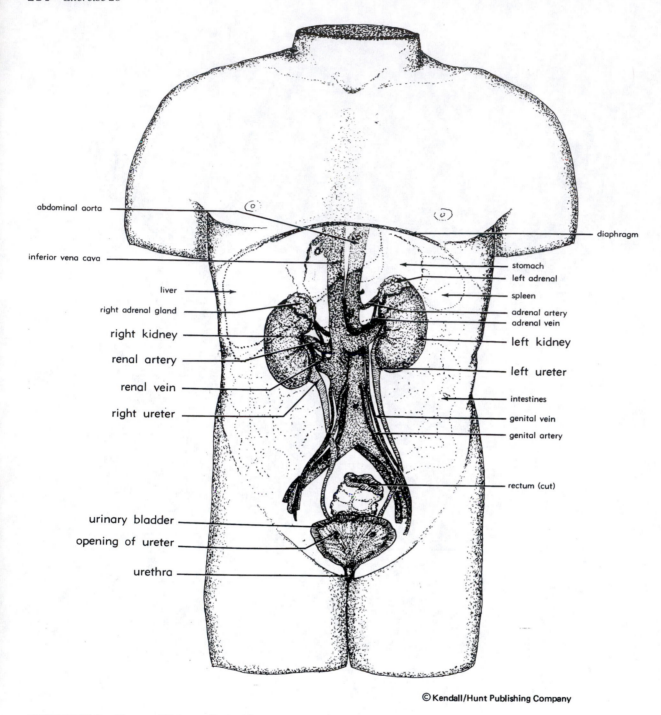

abdominal aorta

inferior vena cava

liver

right adrenal gland

right kidney

renal artery

renal vein

right ureter

urinary bladder

opening of ureter

urethra

diaphragm

stomach

left adrenal

spleen

adrenal artery

adrenal vein

left kidney

left ureter

intestines

genital vein

genital artery

rectum (cut)

FIGURE 25.3   Human (Urinary System)

Human Urinary System

Using a human torso model and preserved sheep kidney, identify similar renal structures in the mammalian body. Paired kidneys, paired ureters and a single urinary bladder are easily identified. Notice the color differences between the kidney center **medulla** and the outer **cortex** regions. Larger renal tubules drain into the central **pelvis** area and exit the **ureter**. Also, note the **renal artery** and **vein** adjacent to the ureter (Figure 25.3, 25.4, Plate 25.1).

The functional unit of excretion in the kidney is the **nephron**. There are approximately one million per kidney. The nephron consists of the **renal corpuscle** enclosed by a **Bowman's capsule**. Obtain a prepared slide of the kidney. Among the tissue, rounded tubules are lined with **simple cuboidal epithelium** allowing for movement of certain materials to

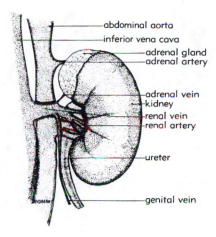

Figure A.   External View

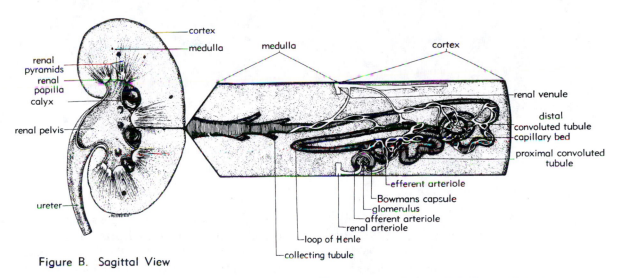

Figure B.   Sagittal View

Figure C.   Diagrammatic Section Uriniferous Tubule

FIGURE 25.4   Human Kidney

be secreted or reabsorbed along their length. Find the large, circular renal corpuscles surrounded by a thin, **simple squamous** layer of epithelium (Figure 25.4, Plate 25.2). The globular material within the capsule is the bed of capillaries or **glomerulus**. As blood flows through the capillary bed, it is filtered and collected in the renal tubules. These become larger in size, drain into the pelvic area of the kidney and the urine exits through a single ureter.

Human Reproductive System

Using a human torso model or Figures 25.5 and 25.6, identify the human reproductive organs. In the human male, paired testes are supported outside the body by the **scrotal sac**. The **seminiferous tubules** within the testes produce the sperm and collect into the **vas deferens**. These tubes progress into the pelvic cavity and transport the sperm through

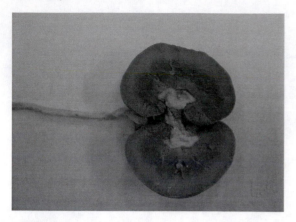

PLATE 25.1   Kidney

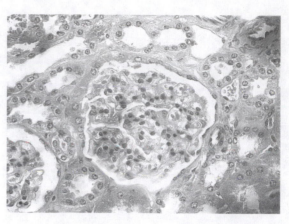

PLATE 25.2   Kidney (Histology)

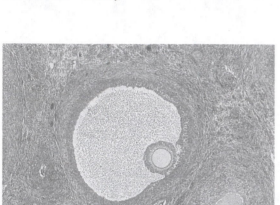

PLATE 25.3 A   Graafian follicle

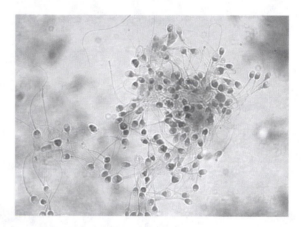

PLATE 25.3 B   Sperm

the **ejaculatory duct**. Along the path of the ejaculatory duct, additional secretions from the **Cowper's gland** and **prostate gland** are added to the sperm. The combined fluid, **semen** is ejaculated out through the urethra in the penis.

In the female, paired **ovaries** contain the developing eggs. At ovulation, one moves into the **Fallopian tubes** leading into the **uterus**. The **vagina**, or birth canal leads to the external surface surrounded by the **labium minor** and major. Superior to this opening is the urethral opening.

Review the histological preparations of both reproductive structures. In the mammalian ovary, locate the **Graafian follicles** and find one with a mature **oocyte**. A protective layer of cells, the **corona radiata**, surrounds this female sex cell. The rest of the follicle is a fluid filled **antrum** (Plate 25.3 A).

Locate mature mammalian sperm on a prepared slide. If several different slides are available, note the differences in the **head** region. Human sperm heads are oval-shaped while rat sperm have a "fish-hook" head. The flagellated tail propels the sperm within the liquid matrix of the semen (Plate 25.3 B).

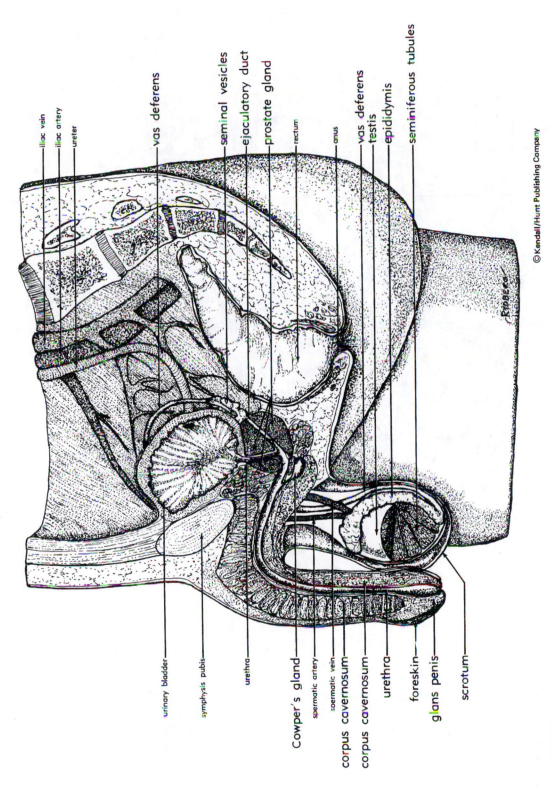

iliac vein
iliac artery
ureter
vas deferens
seminal vesicles
ejaculatory duct
prostate gland
rectum
anus
vas deferens
testis
epididymis
seminiferous tubules

urinary bladder
symphysis pubis
urethra
Cowper's gland
spermatic artery
spermatic vein
corpus cavernosum
corpus cavernosum
urethra
foreskin
glans penis
scrotum

FIGURE 25.5 Human (Male Reproductive System)

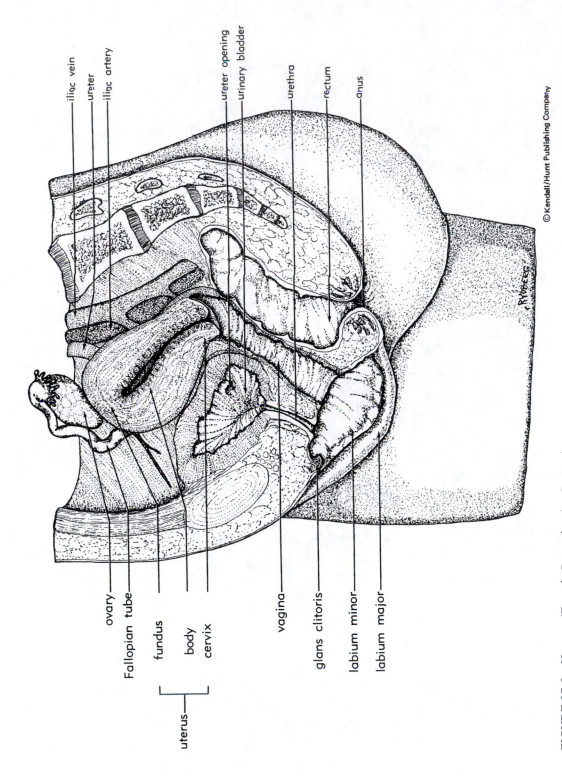

iliac vein
ureter
iliac artery
ureter opening
urinary bladder
urethra
rectum
anus

ovary
Fallopian tube
fundus
body
cervix
uterus
vagina
glans clitoris
labium minor
labium major

© Kendall/Hunt Publishing Company

FIGURE 25.6   Human (Female Reproductive System)

Questions

1. Trace the path of urine from a kidney to the outside.

2. How is urine formed by the nephron?

3. Trace the path of sperm from the testis to the outside.

4. Trace the path of ova from the ovary to the outside.

5. Identify the major differences between amphibian and mammalian reproduction.

EXERCISE 26　The Frog: Respiratory System

Learning Objectives

- ✔ Identify the structures of the frog respiratory system and their functions.
- ✔ Identify the structures of the human respiratory system and their functions.
- ✔ Explain what changes occur to the respiratory system in adapting from an aquatic to a terrestrial environment.

Introduction

The function of the respiratory system is to obtain oxygen from the atmosphere and to eliminate carbon dioxide from the body. Oxygen is necessary for **aerobic respiration** carried on in the mitochondria of the cell. In this process, the breakdown of sugar ($C_6H_{12}O_6$) molecules releases energy to form adenosine triphosphate (ATP). Oxygen entering the cell is combined with hydrogens released from the sugar to form **metabolic water**. Since this water cannot be used to further energy production, it is usually considered waste and excessive amounts are removed by the kidneys (excretion). The carbons from the sugar molecules are removed from the body as CO_2 gas through a respiratory organ; namely, the lungs in air-breathing vertebrates.

Frog Respiratory System

Although the amphibians were the first terrestrial forms to have lungs, they are not very efficient. Adult frogs have no muscular diaphragm or ribs to move a thoracic cage. To breathe, they must inflate the mouth and force air into the lungs. The movement of air into the mouth occurs when the frog closes the glottis and depresses the floor of the mouth. This drops the pressure in the mouth below that of the atmosphere and air moves into the mouth. By lifting the floor of the mouth and closing the nares, the air pressure in the mouth exceeds that of the lungs. When the glottis is opened, air is forced into the lungs. The frog respiratory system is known as a **positive pressure** system since a greater pressure than that of the atmosphere is created in the mouth.

　Move the heart of your frog to either side to expose the lungs. Dissect a lung and notice the small sacs or **alveoli** around the lining. Locate the **bronchus** (pl; bronchi) that connects each lung to the **larynx** (Figure 26.1). Because frogs have no neck, they have no long windpipe or **trachea**.

　To aid the lungs, the frog also obtains oxygen by other methods including: (1) through capillaries in the roof of the mouth, and (2) through capillaries in the skin (**cutaneous respiration**). The skin provides a constant source of oxygen and is the only source while the frog is hibernating. When active, the lining of the mouth provides an additional amount of oxygen. Depending on the frog's activity, it can increase or decrease the rate of gas exchange between the mouth and the atmosphere.

　Notice the thickness of the skin still remaining on the frog's head. In many places, the arteries are so close to the surface of the skin that the red latex used to inject arteries will color the skin pink. Likewise, the lining of the mouth will be pink.

Human Respiratory System

Since the lungs respond like a balloon to changes in internal and external pressure, a mechanical process is employed to move air into and out of the lung. In mammals, this change in the air pressure of the lung is effected by the diaphragm muscle which separates the thoracic cavity from the abdominal cavity. The diaphragm and chest muscles contract to enlarge the chest cavity which causes the pressure in the lungs (thoracic cavity) to drop below that of the atmosphere. Since air will move from a region of high to low pressure, this drop in pressure causes the movement of air into lungs. Thus, the mammalian respiratory system is known as a **negative pressure** system. When the diaphragm relaxes and the air in the lung is compressed, pressure in the lung exceeds that of the atmosphere and air rushes out. Again, the air is moving from a region of higher to lower pressure. Thus, air is exchanged between the external environment of the mammal and that of the respiratory organs.

　Using Figure 26.2 or a human torso model, identify the parts of the human respiratory system. Air enters the **nostrils** and **nasal cavity** and moves to the back of the throat by way of the **nasopharynx**. As air moves down into the lower

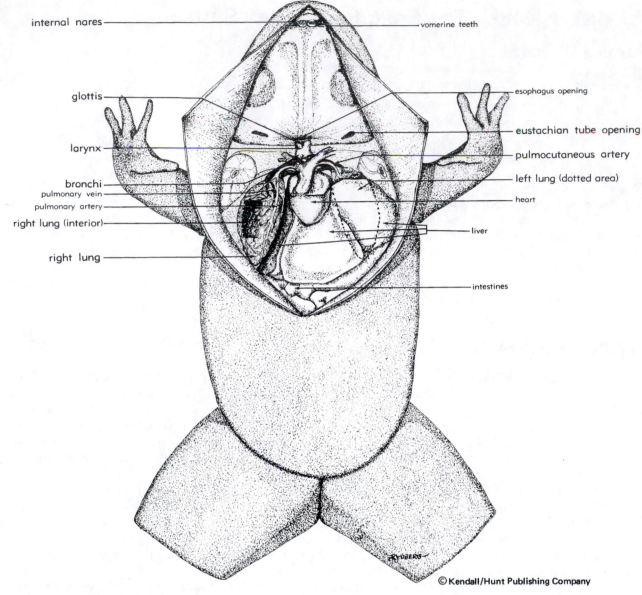

internal nares — — vomerine teeth

glottis — — esophagus opening

— eustachian tube opening

larynx — — pulmocutaneous artery

— left lung (dotted area)

bronchi —
pulmonary vein —
pulmonary artery — — heart

right lung (interior) —

right lung — — liver

— intestines

© Kendall/Hunt Publishing Company

FIGURE 26.1   Frog (Respiratory System)

portion of the pharynx, it is sent through the **larynx** or "voice box." Look for the **thyroid cartilage** or Adam's apple that identifies the position of the larynx. Internally, the **vocal cords** are involved in sound production. The windpipe, or **trachea**, has C-shaped rings of cartilage protecting it. Splitting into two **bronchi**, the tubes enter the **lung** proper and branch into smaller **bronchioles**. Actual exchange of gases to and from the blood occurs at the terminal **alveolar sacs**.

Action of the Diaphragm

The action of the diaphragm muscle can be demonstrated by an apparatus, or **bell jar**, with a rubber plunger fitted to the lid (Plate 26.1). A "Y" glass tubing with two balloons have been fitted to the lid and is open to the atmosphere at the top. Complete the following experiments:

a. Move the plunger down. What happens? Why?

b. Move the plunger up into the jar. What happens? Why?

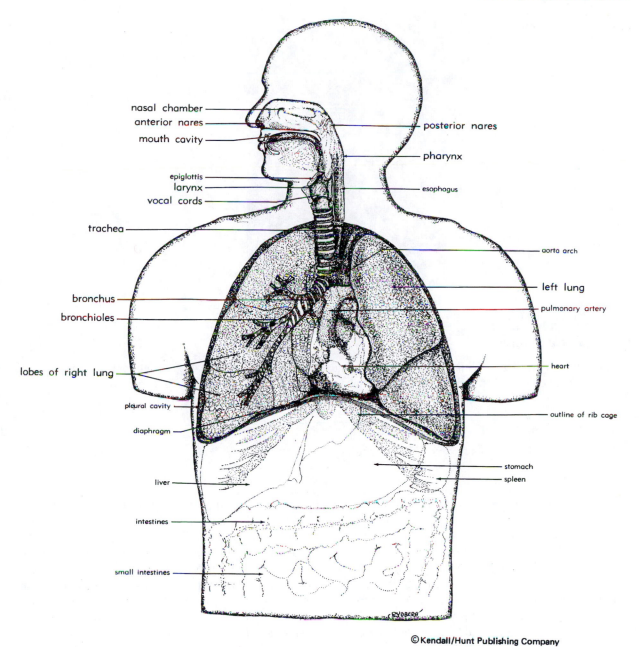

nasal chamber

anterior nares

mouth cavity

posterior nares

pharynx

epiglottis

larynx

vocal cords

esophagus

trachea

aorta arch

left lung

bronchus

pulmonary artery

bronchioles

lobes of right lung

heart

pleural cavity

outline of rib cage

diaphragm

stomach

spleen

liver

intestines

small intestines

FIGURE 26.2   Human (Respiratory System)

Compare these different parts with structures of the human body.

opening at the top of bell jar _____

"Y" tube _____

balloons _____

bell jar _____

plunger _____

PLATE 26.1   Bell Jar

Metabolic Waste Product

Obtain an Erlenmeyer flask and fill it approximately one-half full of water. Add three drops of **phenol red**, a pH indicator. (Notice the pink color). With a clean soda straw blow your breath into this solution. Continue until there is a noticeable color change.

    a. Write the chemical formula that describes this reaction.

    b. Does this prove that CO_2 is given off by the lungs?

Wet or Dry Spirometer

The **spirometer** is an apparatus (Plate 26.2) to measure **vital capacity**, the maximum volume of air that can be expelled with one breath. Vital capacity consists of three volumes of air. The normal volume of air that is inhaled and exhaled is the **resting tidal volume** (@ 0.5 l). After a normal inhalation, an additional amount called the **inspiratory reserve volume** (@ 1.5 l) can be inhaled. After a normal exhalation, an additional amount called the **expiratory reserve volume** (@ 1.5 l) can be exhaled. Males have an average vital capacity of 4.0 liters while females have an average of 3.0 liters. Consequently, the vital capacity averages about 3.5 liters for the total population. Even though a significant volume of air can be forcefully exchanged with the environment, there is a volume of air that cannot be expelled from the lungs. This volume of air is known as **residual volume** (@ 1.5 l) (Figure 26.3).

    Vital capacity can be measured by using either a dry or wet **spirometer.** Complete the following steps to reach a personal value:

1. Obtain a clean mouthpiece and fit it to the instrument.
2. Take as deep a breath as possible and then blow into the spirometer. (Do not blow **too** hard or air will escape around the mouthpiece.)
3. Record the amount of air expelled._____
4. Compare this value to the posted chart for vital capacity. (The amount of air expelled is given in cubic centimeters and inches across the top and bottom of the chart. The standing height is in inches and centimeters on the side of the chart.)
5. Find the value that corresponds as close as possible to your standing height in either inches or centimeters.
6. Locate the vertical column with your vital capacity in cubic centimeters or cubic inches.
7. Follow this column with your vital capacity until it intersects the row with your standing height.
8. The value in the square where the two columns intersect is the percentage of the vital capacity for the **average** person of that height.

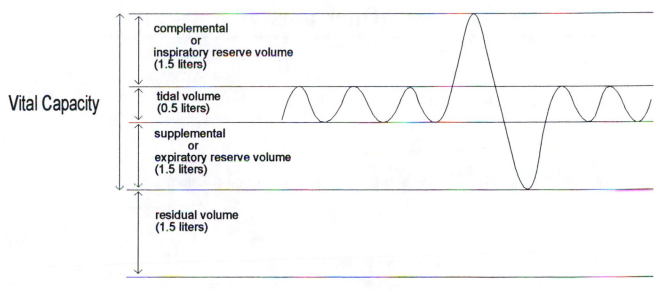

FIGURE 26.3   Chart of vital capacity

PLATE 26.2   Spirometer

9. Now expel a normal breath of air into the spirometer. How much air did you expel _____?

10. If air contains approximately 21% oxygen, 79% nitrogen, and trace amounts of other gases, determine the amount of **oxygen** in a:

normal exhalation _____

forced exhalation _____

Questions

1. What is the function of the epiglottis in man?

2. What is the advantage and disadvantage of having cutaneous respiration as in the frog?

3. Where and how is oxygen transported within the body?

4. What is the location and action of the human diaphragm?

5. Explain the difference between breathing and cellular respiration.

6. How does a negative pressure system differ from that of a positive pressure system?

7. In what structures of the lung does gas exchange occur between the air in the lung and the blood?

EXERCISE 27   The Frog: Circulatory System

Learning Objectives

✔ Identify the various structures of the circulatory system and give their function(s).
✔ Describe the path that blood takes through the heart.
✔ Describe the path that blood takes from the heart to make a complete circuit through the body.

Introduction

The circulatory system is the body's transportation medium. It consists of a **heart**, blood **vessels** and the **blood** that flows through them. The pumping action of the muscular heart moves the blood throughout the body transporting gases, food, waste, hormones and other materials. The blood vessels include: **arteries** and **arterioles** which carry blood away from the heart; **veins** and **venules** which carry blood to the heart and **capillaries**, the connecting tubes between the two. In air-breathing animals, the circulatory system is subdivided into two major pathways called **systemic** and **pulmonary**. The systemic circulation involves the flow of blood between the heart and the body, while the pulmonary includes the flow of blood between the heart and the lungs. At the lungs, the exchange of materials occurs at the capillaries associated with alveoli. Generally, oxygen gas is picked up by the blood and carbon dioxide is sent into the lungs for exhalation. Oxygenated blood returning to the heart is then pumped to the entire body where exchange of materials occurs at the capillaries adjacent to body cells. Blood is cleansed of chemical impurities when it enters the kidney; a part of the excretory system.

Frog Circulation

Frogs have a three-chambered heart but form a two-circuit circulatory system. The blood goes from the heart to the lungs (pulmonary), back to the heart, and then out to the various body systems (systemic). In the frog; however, the three-chambered heart does not completely separate deoxygenated from oxygenated blood. The two top chambers or **atria** receive the blood (right atrium from the body and left atrium from the lungs) and the single, large **ventricle** pumps the slightly mixed blood out to the body.

Obtain your frog from the container. Note that the vascular system of the frog has been injected with a rubber latex; red in arteries and blue in veins. Fold back the ventral skin flaps and expose the heart in its protected chest region. Identify the three heart chambers. Locate the tubular **conus arteriosus** which exits the heart and branches into two large arteries, the **truncus arteriosus** or **aortic arches**.

Arteries

Trace the red aortic arches from the heart, cleaning away debris and connective tissue. Three branches lead from each. These are the: (1) **carotid arch** leading to the head, (2) **systemic arch** leading to the gut region, and (3) **pulmocutaneous arch** feeding the skin and lungs. The common carotid arch divides into the **external carotid** (to floor of mouth) and **internal carotid** (dorsal roof of mouth). The pulmocutaneous branches into the **pulmonary** (to lungs) and **cutaneous** (to skin on dorsal surface). The systemic arteries have major branches; the **subclavian** and **brachial** which progress into the arm region. Trace the systemic on one side and note that the two branches come together to form a single vessel, the **dorsal aorta** in the gut region under the liver and stomach. The dorsal aorta follows along the vertebral column to the pelvic area (Figure 27.1).

At the junction of the two systemic arteries, the **coeliacomesenteric** artery branches off to the liver, stomach, intestine, and spleen. In the pelvic area, the dorsal aorta forks into two common **iliac arteries** which supplies blood to the urinary bladder, the cloaca and the skin of the thigh. The main branch of the iliac continues down into the deep thigh as the **sciatic artery**. Locate the **spleen** in the mesentery near the small intestine. The dark, rounded organ stores red blood cells and removes worn out red blood cells.

Veins

Venous blood flows toward the heart. Name each blue vein beginning with a hind limb (Figure 27.2). Trace the ventral abdominal vein toward the groin area. Move the urinary bladder aside and follow the right and left **pelvic** in the groin

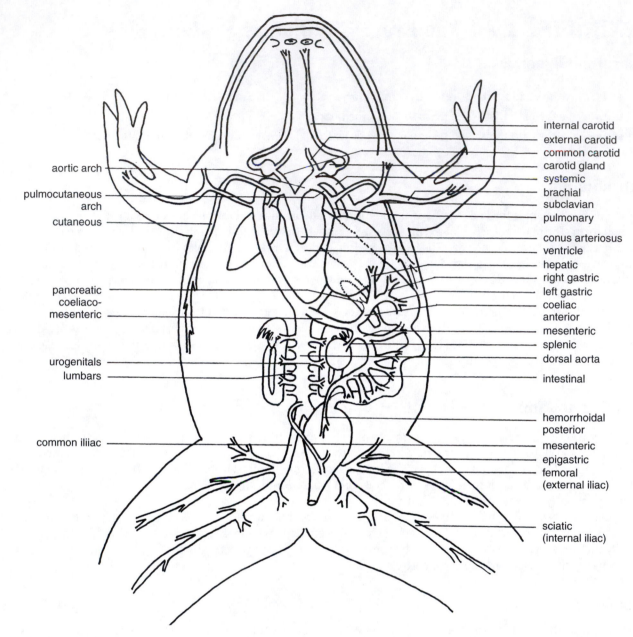

FIGURE 27.1 Frog arterial system

Labels (left side, top to bottom): aortic arch, pulmocutaneous arch, cutaneous, pancreatic coeliaco-mesenteric, urogenitals, lumbars, common iliac

Labels (right side, top to bottom): internal carotid, external carotid, common carotid, carotid gland, systemic, brachial, subclavian, pulmonary, conus arteriosus, ventricle, hepatic, right gastric, left gastric, coeliac, anterior, mesenteric, splenic, dorsal aorta, intestinal, hemorrhoidal, posterior, mesenteric, epigastric, femoral (external iliac), sciatic (internal iliac)

area. Follow one of the pelvic veins to find the **femoral** vein on the anterior side of the thigh. Turn to the posterior side of the thigh and find the **sciatic** vein (generally visible between the triceps femoris and biceps femoris muscles).

At the convergence of the pelvic, sciatic and femoral veins the **renal portal vein** moves up along the lateral edge of each kidney. **Renal** veins extend as parallel, small branches from the midline between the two kidneys. They leave and form the **inferior** or **posterior vena cava**, which drains into the **sinus venosus**. This enlarged vein enters into the right atrium. **Hepatic** veins from the liver move blood into the inferior **vena cava** just behind the sinus venosus. Veins from the stomach, intestine, pancreas, and spleen form the **hepatic portal system** which empties into the liver.

Human Heart

The human heart is a four chambered organ. The separation of the ventricle completely separates the oxygenated blood from the deoxygenated blood. Study a mammal heart (pig, cow, or sheep) and identify significant structures on the external surface and internal chambers (Plate 27.1).

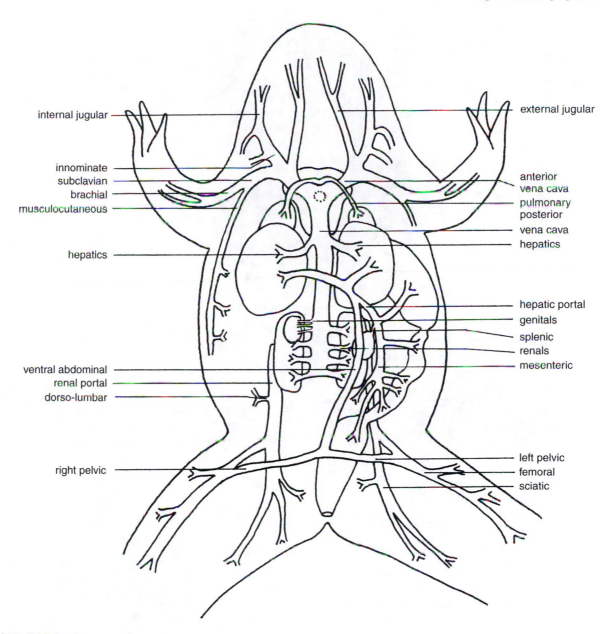

internal jugular

external jugular

innominate
subclavian
brachial
musculocutaneous

anterior
vena cava
pulmonary
posterior
vena cava
hepatics

hepatics

hepatic portal
genitals
splenic
renals
mesenteric

ventral abdominal
renal portal
dorso-lumbar

right pelvic

left pelvic
femoral
sciatic

FIGURE 27.2 Frog venous system

Locate the **coronary** blood vessels on the external surface. These arteries and veins supply the heart tissue itself. In an anterior view, the **aorta** is the large vessel extending dorsal to the heart. Slightly anterior to the aorta, the large **pulmonary trunk** branches left and right. On the posterior view, the **posterior** and **inferior vena cava** may be visible entering the atrium (Figure 27.3).

Open the heart to expose the inner chambers. Orient the heart so that the large left ventricle is to your right. Find the same structures previously identified: vena cava, right and left atria, right and left ventricle. Note the size of the wall of each ventricle and thickness of the **interventricular septum**. Identify the valves which control the flow of blood through the heart. Between the right atrium and right ventricle lies the **tricuspid** (atrioventricular) valve. Between the left atrium and left ventricle lies the **bicuspid** (atrioventricular) valve, or **mitral** valve. Both of these are easily recognized by the fibrous **chordae tendineae** attached to **papillary muscles** on the ventricular walls. The aortic **semilunar** valve sits in the aorta and controls backflush of blood between the aorta and the left ventricle. Using Figures 27.4 and 27.5, note the similarity in name and location of the major blood vessels in the human body to those of the frog.

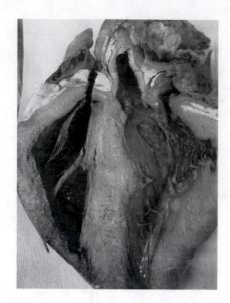

PLATE 27.1 Mammalian Heart

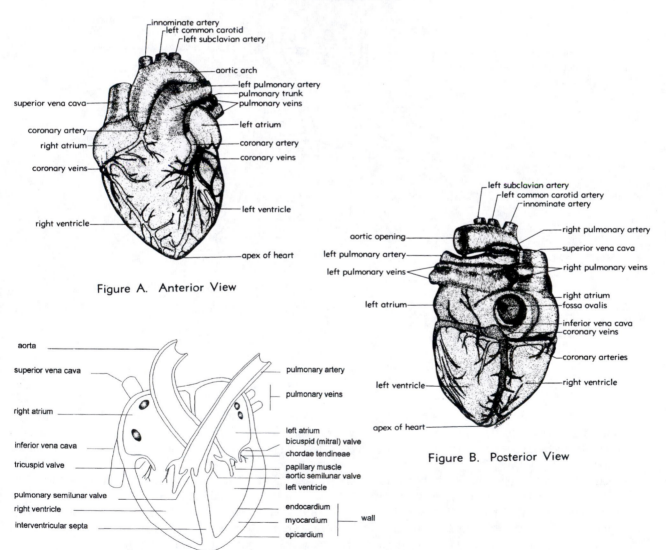

innominate artery
left common carotid
left subclavian artery
aortic arch
left pulmonary artery
pulmonary trunk
pulmonary veins
superior vena cava
left atrium
coronary artery
coronary artery
right atrium
coronary veins
coronary veins
left ventricle
right ventricle
apex of heart

Figure A. Anterior View

left subclavian artery
left common carotid artery
innominate artery
aortic opening
right pulmonary artery
left pulmonary artery
superior vena cava
left pulmonary veins
right pulmonary veins
right atrium
fossa ovalis
left atrium
inferior vena cava
coronary veins
coronary arteries
left ventricle
right ventricle
apex of heart

Figure B. Posterior View

aorta
superior vena cava
pulmonary artery
pulmonary veins
right atrium
inferior vena cava
left atrium
bicuspid (mitral) valve
chordae tendineae
tricuspid valve
papillary muscle
aortic semilunar valve
left ventricle
pulmonary semilunar valve
endocardium
right ventricle
myocardium   wall
interventricular septa
epicardium

Figure C. Sagittal View

FIGURE 27.3 Human Heart

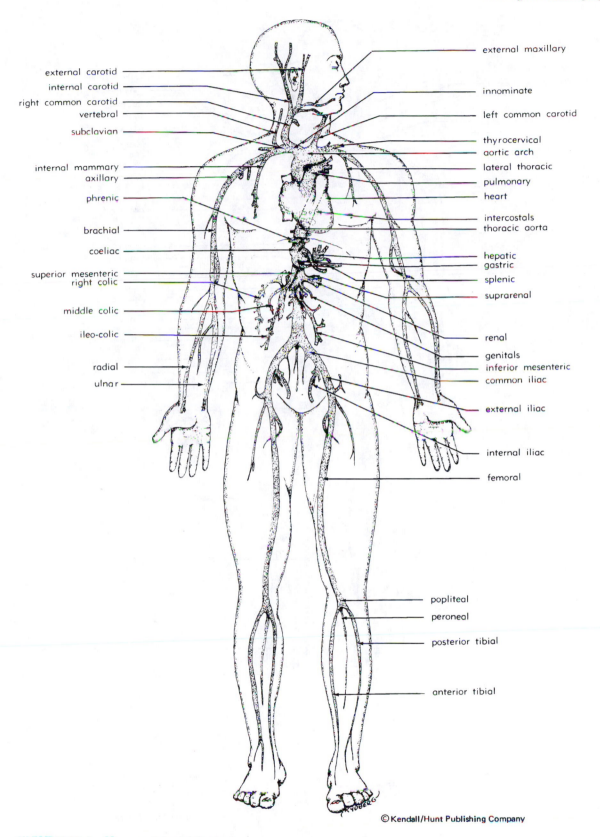

external carotid
internal carotid
right common carotid
vertebral
subclavian

internal mammary
axillary
phrenic

brachial

coeliac

superior mesenteric
right colic

middle colic

ileo-colic

radial

ulnar

external maxillary

innominate

left common carotid

thyrocervical
aortic arch
lateral thoracic
pulmonary
heart
intercostals
thoracic aorta

hepatic
gastric
splenic
suprarenal

renal

genitals
inferior mesenteric
common iliac

external iliac

internal iliac

femoral

popliteal
peroneal

posterior tibial

anterior tibial

FIGURE 27.4 Human Arterial System

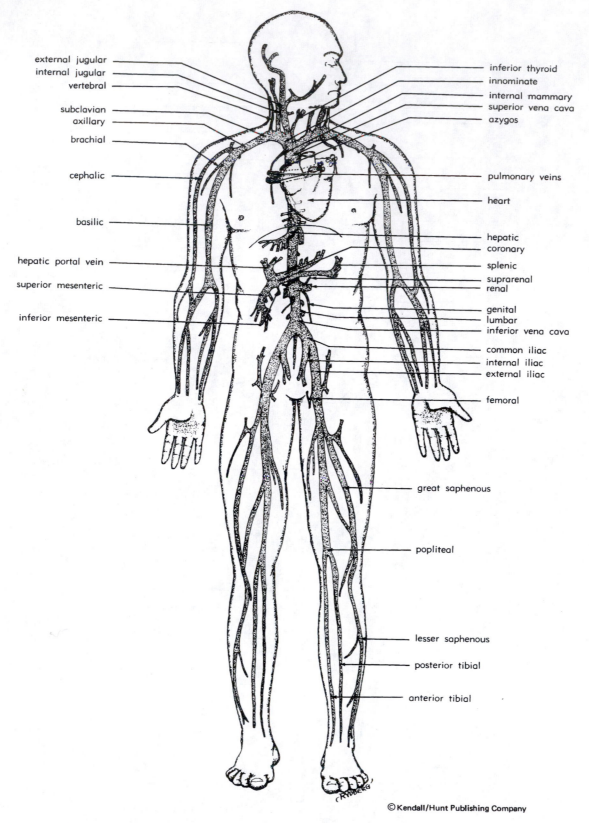

external jugular
internal jugular
vertebral

subclavian
axillary

brachial

cephalic

basilic

hepatic portal vein

superior mesenteric

inferior mesenteric

inferior thyroid
innominate
internal mammary
superior vena cava
azygos

pulmonary veins

heart

hepatic
coronary
splenic
suprarenal
renal

genital
lumbar
inferior vena cava

common iliac
internal iliac
external iliac

femoral

great saphenous

popliteal

lesser saphenous

posterior tibial

anterior tibial

FIGURE 27.5 Human Venous System

Questions

1. What is the purpose of the:
   a. Ventricle?

   b. Atria?

2. Does oxygenated and deoxygenated blood normally mix in the:
   a. Frog ventricle?

   b. Human ventricle?

3. Trace blood from the conus arteriosus to the stomach and back to the conus arteriosus.

4. Trace the path of blood through the human heart.

5. How does a closed circulatory system differ from that of an open circulatory system?

6. In what type of vessel does the exchange of nutrients and waste occur?

7. What is the cause of the heart sounds, "lub-dub?"

Blood and Blood Groups

Learning Objectives

✔ Explain how combinations of red blood cell antigens produce different blood types.

✔ Explain the role of antibodies in their defense of the body.

✔ Use a sphygmomanometer to determine blood pressure.

Introduction

Blood is composed of plasma and formed elements. Plasma is the water-based, liquid portion and the formed elements are the cellular components. Earlier in the last century, medical researchers found that blood could be typed and those with the same blood type could exchange blood if necessary.

The various blood types are a reflection of the differences that exist between individuals. Genetically, we all differ in the types of proteins on the surface of our cells. These proteins are known as **antigens** because they stimulate the production of **antibodies** (proteins). Antibodies are produced by the individual's immune system when an antigen is introduced into their body. Antibodies complex with the antigens and neutralize them. Antibodies may act as **antitoxins** by combining with a **toxin** (poison), such as those produced by certain bacteria or snake venom, to neutralize the toxin. The antibody may be a **lysin**, an antibody that complexes with an antigen on a cell surface and causes the cell to lyse or burst. Or, an antibody may serve as an **agglutinin**. This is an antibody that attaches to an antigen on two or more cell membranes simultaneously and **agglutinates** (clumps) the cells together. Regardless of the mode of action that an antibody has, a potentially harmful substance can be rendered harmless. In this way, our immune system can distinguish between "self" and "nonself."

Antibodies may be of two types, **acquired** or **natural**. Acquired antibodies are produced upon response to an antigen that has been *introduced* into the body. For example, tetanus toxin, viruses, or bacteria. Natural antibodies are produced in the *absence* of the appropriate antigen and are available in the event the antigen is introduced into the body. An example of this is the antibody production of the ABO blood groups.

Red blood cells have proteins on the surface of the cell just as other types of cells do. If the wrong type of blood is given, then these proteins will be recognized as alien and the body will attempt to neutralize these cells with antibodies. Consequently, when blood transfusions are given, it is important to know the blood type of the donor and the recipient. If blood of the wrong type is given, then serious complications can result.

Blood Groups

Two types of antigens, A and B, may be found on the surface of a red blood cell (Figure 28.1, Table 28.1). The presence or absence of these antigens results in four groups or blood types. **Genotypes** indicate the possible genetic combinations in an individual's chromosomal DNA. In addition, another protein known as the **Rh** factor may or may not be present (Table 28.2). As a result of the different number of possible combinations of the antigens, the following blood types are now recognized:

TABLE 28.1   Blood Types

Type	% in Population	Antigen on RBC Surface	Antibody in Plasma	Genotype
A	42	A	b	AA, AO
B	10	B	a	BB, BO
AB	4	A, B	—	AB
O	44	—	a, b	OO

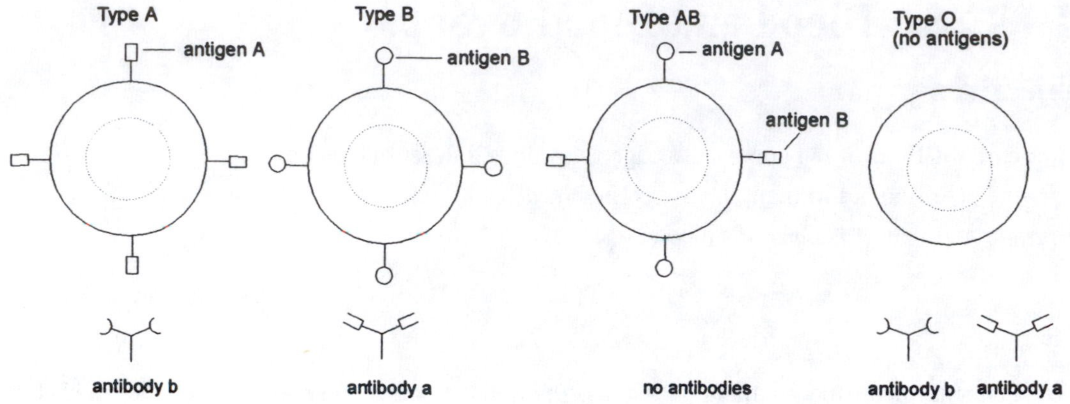

A. Different Blood Types. Notice that the antigens are embedded in the cell membrane and the antibodies are dissolved in the plasma.

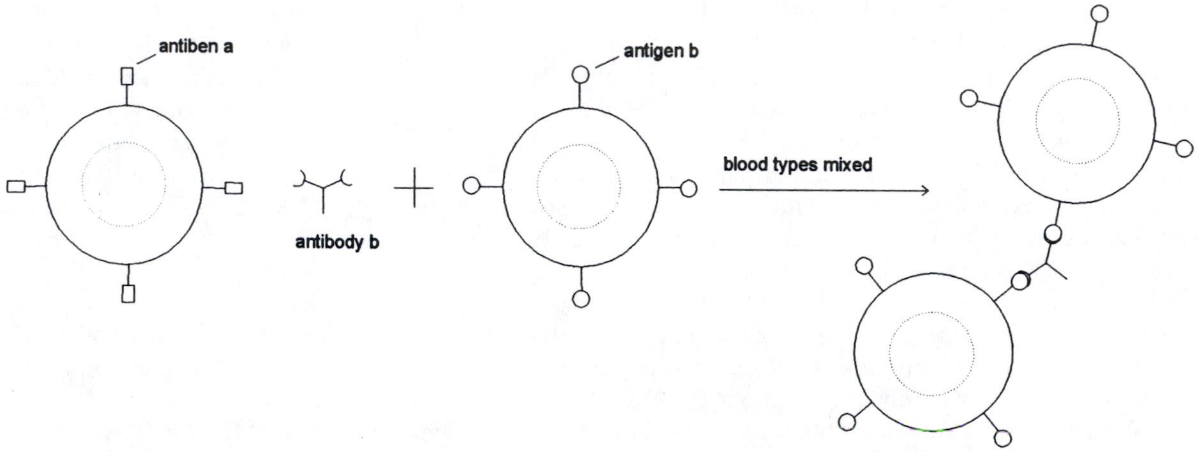

B. An example of the Agglutination of Blood. Type B blood has been added to type A blood. The antibodies in the type A blood have agglutinated the type B blood cells..

FIGURE 28.1   Blood typing.

Any of the above blood types may be positive or negative for the Rh factor.

TABLE 28.2   Rh Factor

Type	% in Population	Antigen on RBC Surface	Antibody in Plasma	Genotype
+	85	D	—	Dd, DD
−	15	—	d	dd

Identification of Blood Type

For safety reasons, human blood will not be used for typing. However, several blood grouping kits are available that accurately portray the procedure and findings. Complete the following steps:

1. Clean two blank slides.
2. With a wax pencil, draw a line down the middle of one slide. Mark −a on one half and −b on the other half of the slide.
3. Mark the second slide with the wax pencil as −Rh.

4. Apply one drop of "fake" blood on each end of the marked slide and one drop on the second slide.

5. To the anti-a portion of the slide, add a drop of anti-a solution near the drop of blood (DO NOT PLACE THE TIP OF THE ANTI-SERA DIRECTLY INTO THE BLOOD DROP).

6. To the anti-b portion of the opposite end, add a drop of anti-b solution near the drop of blood.

7. Mix the drops of blood and solution. Toothpicks are excellent for stirring (Note: Do **not** use the same toothpick for both solutions. Do **not** allow the adjacent two solutions to run into each other).

8. Observe the solutions for clumping or agglutination. Antibody-a agglutinates antigen-A and antibody-b agglutinates antigen-B. What is your blood group?_____

9. Using another toothpick, stir the blood and anti-Rh solution together. Is your blood type, Rh+ or Rh−?_____

Identification of Blood Corpuscles

Obtain a prepared slide of human blood stained with Wright's stain. Review the histology of blood tissue from Exercise 3. The majority of cells will be **erythrocytes** or red blood cells (Figure 3.6). They are biconcave discs without a nucleus and are stained a faint red or pink. Move the slide about and find some **leucocytes** or white blood cells These cells are stained a bluish color and are larger than a RBC. Leucocytes are of two general types, **granulocytes** and **agranulocytes.** Granulocytes have granules in the cytoplasm with the nucleus having two or more lobes. Agranulocytes lack granules in the cytoplasm. The nuclei of agranulocytes are either round or kidney-shaped. The platelets, or **thrombocytes,** appear as small particles among the other cellular pieces.

Determination of Blood Pressure by Sphygmomanometer

The sphygmomanometer is an instrument for measuring blood pressure. An inflatable cuff and readable dial are provided with each instrument. Follow the procedure below and listen carefully to determine a fellow student's blood pressure:

(**Caution:** Do not leave an inflated cuff on the arm for prolonged time periods)

1. Wrap the cuff around the upper part of the arm putting pressure on the brachial artery.

2. Place a stethoscope over the artery at the bend of your arm.

3. Tighten the regulator screw on the bulb end of the cuff.

4. Inflate the cuff to about 150 mm. of mercury on the dial.

5. Listen with the stethoscope while letting air escape **slowly** from the cuff. When a sound is evident, this is the **systolic** sound or pressure. This is the maximum pressure produced by the heart. Read the amount of mm. of mercury at this point.

6. Gradually release more air until you can no longer hear any sound. Read the amount of mm. of mercury at this point. This is the minimum pressure or **diastolic** pressure produced by the heart.

Systolic sound is lower in pitch and is of longer duration than the louder diastolic sound. Systolic sound occurs mainly with the closing of the cuspid valves during ventricular contraction. Diastolic sound occurs mainly from the closing of the semilunar valves. If time permits, determine the blood pressure of other group members.

Questions

1. What is the composition of blood tissue?

2. Give several functions that blood serves.

3. Why can you visualize the different blood types after applying antisera?

4. Could blood typing be used to establish paternity in all cases? Explain.

5. What is the function of the following formed elements:

   Erythrocytes?

   Neutrophils?

   Basophils?

   Eosinophils?

   Lymphocytes?

   Monocytes?

   Platelets?

EXERCISE 29 Central Nervous System and Sense Organs

Learning Objectives

✔ Identify the different regions of the brain and their functions.
✔ Identify the structures of the eye and how vision is detected.
✔ Identify the structures of the ear and how hearing and balance are detected.

Introduction

Along with the endocrine system, the nervous system coordinates the different organ systems of the body in their activities. It is composed of two major divisions. The first is the **central** nervous system which consists of the **brain** and **spinal cord**. In the frog and other vertebrates the central nervous system is located dorsally in the body. The second is the **peripheral** nervous system which is composed of the **autonomic** nervous system together with the **spinal** and **cranial nerves**.

Nervous System of the Frog

Obtain your frog and place it in the dissecting pan. With scalpel, scissors and forceps, clean the head region, removing any skin or tissue covering the skull. Begin exposing the brain by scraping or clipping away bits of the bony cranium. Be careful in the dissection and do not damage the brain. Expose the entire portion of the brain and about one inch of the spinal cord. The **meninges** is the pigmented membranes that surround the nervous system. A tough, outer membrane, the **dura mater** normally remains with the cranium. A more delicate membrane, the **pia mater**, is internal and usually remains with the brain and spinal cord when the central nervous system is dissected (Figure 29.1).

Identify parts of the **forebrain** on the dorsal surface. They include the **olfactory lobes**, a paired extension used for the sensation of smell and two enlarged **cerebral hemispheres**. Two large, round, **optic lobes**, form the **midbrain** area. In the **hindbrain**, a small transverse **cerebellum** lies posterior to the optic lobes and continues as the **medulla**. Tapering into the **spinal cord**, the nervous system continues into the vertebral column.

Remove the entire exposed portion of the central nervous system by lifting the posterior portion of the spinal cord with forceps. Be extremely careful! Continue to lift, snipping each nerve connection. Sever the anterior portion of the olfactory lobes and place the entire brain in a shallow dish ventral side up.

Again, note the olfactory bulbs and cerebral hemispheres in the anterior portion. The **optic nerves** cross at the **optic chiasma** on the ventral surface. A broad, bilobed structure, the **infundibulum**, sits posterior to the chiasma and anterior to the large **pituitary gland**. Numerous **cranial nerve** stubs can be seen emerging from both sides of several brain areas. Because the **spinal cord** has been separated, its **spinal nerve** connections remain within the vertebral column.

Nervous System of the Human

Observe a mammalian brain (Ex: sheep, cow,) or a model of the human brain (Figure 29. 2, Plate 29.1 A, B, C). Locate the following structures either internally or on the external (dorsal or ventral) surfaces.

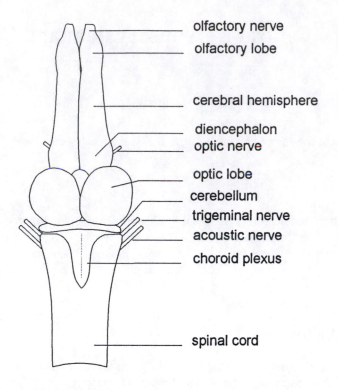

olfactory nerve

olfactory lobe

cerebral hemisphere

diencephalon
optic nerve

optic lobe
cerebellum
trigeminal nerve
acoustic nerve
choroid plexus

spinal cord

A. Dorsal View

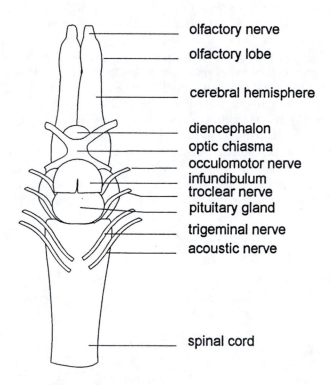

olfactory nerve

olfactory lobe

cerebral hemisphere

diencephalon
optic chiasma
occulomotor nerve
infundibulum
troclear nerve
pituitary gland

trigeminal nerve
acoustic nerve

spinal cord

B. Ventral View

FIGURE 29.1   Frog (Brain and Spinal Cord)

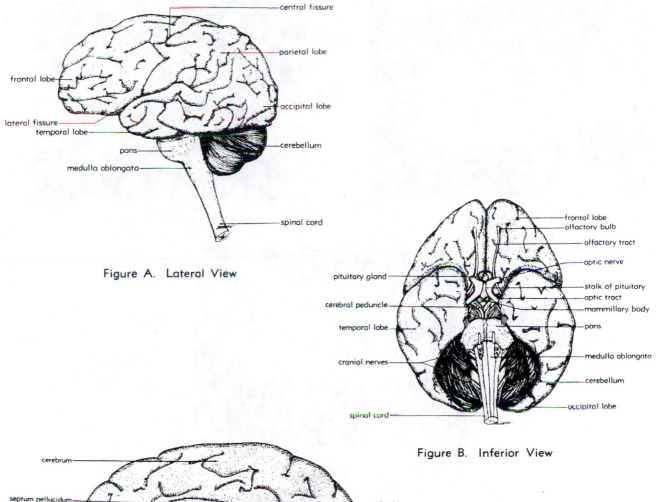

central fissure

parietal lobe

frontal lobe

occipital lobe

lateral fissure
temporal lobe

pons

cerebellum

medulla oblongata

spinal cord

Figure A. Lateral View

frontal lobe
olfactory bulb

olfactory tract

optic nerve

pituitary gland

stalk of pituitary
optic tract
mammillary body

cerebral peduncle

temporal lobe

pons

cranial nerves

medulla oblongata

cerebellum

spinal cord

occipital lobe

Figure B. Inferior View

cerebrum

septum pellucidum

corpus callosum

fornix

thalamus
third ventricle

intermediate mass

optic chiasma

pineal body

pituitary gland

pons

cerebellum

fourth ventricle

medulla oblongata

spinal cord

Figure C. Sagittal View

FIGURE 29.2   Human Brain

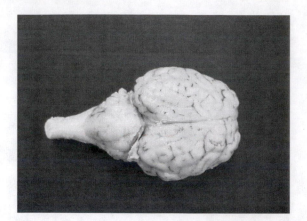

PLATE 29.1 A   Brain (Superior View)

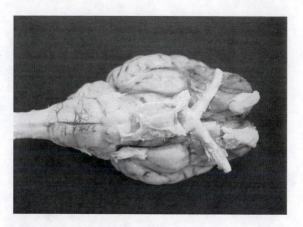

PLATE 29.1 B   Brain (Inferior View)

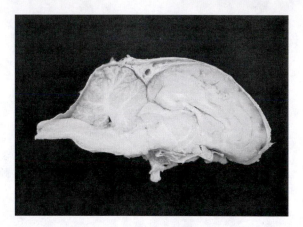

PLATE 29.1 C   Brain (Sagittal View)

Area	Structure	Function
Forebrain	Cerebrum	Higher brain function (sensory + motor)
	Corpus Callosum	Nerve tracts connect left/right cerebral hemispheres
	Thalamus	Relay center
	Hypothalamus endocrine input	Autonomic center,
	Pituitary	Major endocrine gland
Midbrain	Midbrain	Visual + auditory reflex
Hindbrain	Pons respiratory center	Motor + sensory nerve tracts,
	Cerebellum contractions	Coordinate skeletal muscle
	Medulla Oblongata	Motor + sensory nerve tracts, cardiac + respiratory center, reflex centers

Note as well, the olfactory lobes, optic nerves and optic chiasma. In the sagittal cut or view, the third and fourth **ventricles** or chambers are visible beneath the thalamus and at the cerebellum connection, respectively. The inner cerebellum is also very distinctive. The branching effect, or **arbor vitae,** are tracts of white nerve matter. If the protective membrane, the meninges, is still visible, note its toughness.

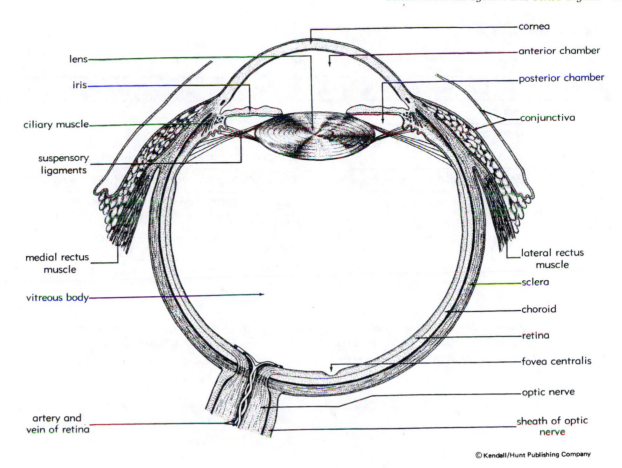

FIGURE 29.3 Human Eye

Place a prepared slide of a cerebral brain section on the dissecting microscope. Note the difference in coloration. The outer dark or **gray matter** covers the cerebral cortex, while the lighter **white matter** fills up the inner brain area. Now place a prepared slide of a spinal cord on the microscope. Note the color reversal and pattern on this section. The white matter is external and the **gray matter** forms into extended **horns**.

Sense Organ: Eye

Use models of the human eye or preserved cow eyes (if available) to locate and identify the parts below (Figure 29.3):

Structure	Function
Cornea	Transmits light
Anterior Chamber	Pressurizes with watery fluid
Iris	Regulates amount of light entering
Pupil	Opening formed by iris
Lens	Shape changes focus image
Vitreous Chamber (Body)	Pressurizes with watery fluid
Retina	Sensory perception of light
Choroid	Supports blood supply
Sclera	Protects eyeball, muscle attachment
Optic Nerve	Transmission of nerve impulse

Sense Organ: Ear

Examine the large model of a human ear (Figure 29.4) to locate and identify the parts below:

Region	Structure	Function
Outer Ear	Pinna	Collects sound waves
	External Auditory Canal	Protection (hairs and wax glands)
	Tympanic Membrane	Vibrates with sound waves, attaches to the malleus bone
Middle Ear	Ear Bones (Malleus, Incus, Stapes)	Vibrate with sound waves, transmit to oval window
	Eustachian Tube (Auditory)	Equalizes pressure between pharynx and middle ear
Inner Ear	Semicircular Canals	Balance perception
	Cochlea	Hearing perception
	Auditory Nerve	Transmit impulses (hearing + balance) to brain

Obtain a prepared slide of the ear cochlea. Note the two large chambers filled with fluid or **perilymph**. The upper **scala vestibuli** and lower **scala tympani** are separated by a membranous chamber or **cochlear duct**. Within this structure lies the specialized spiral organ or **organ of Corti**. Sound vibrations transmitted into fluid wave movements stimulate sensory **hair cells**, resulting in the hearing sensation (Figure 29.4, Plate 29.2).

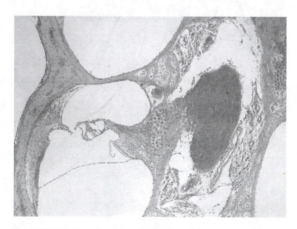

PLATE 29.2   Cochlea (Organ of Corti)

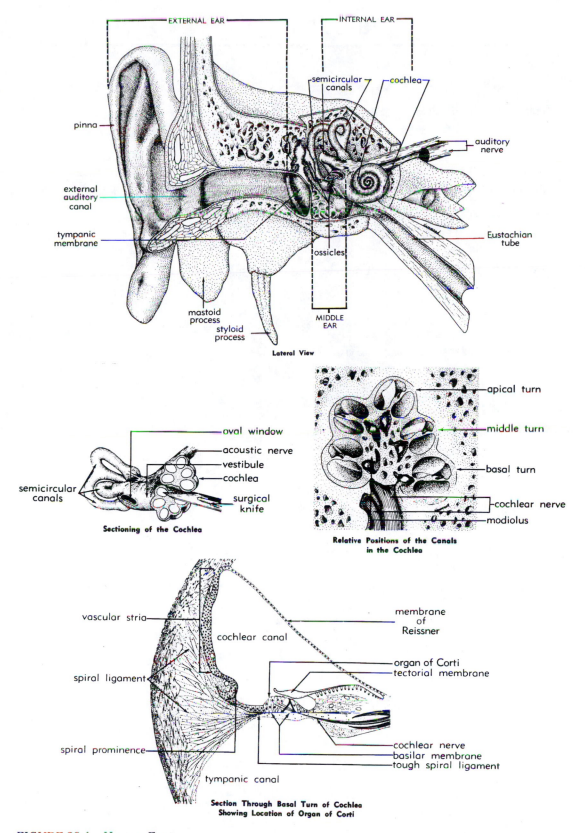

FIGURE 29.4   Human Ear

Questions

1. What major differences are there between the human and frog brain?

2. If a person sustained a hard blow to the back of the head, what physical problems might be expected?

3. Why do strokes to the left side of the brain affect muscles on the right side of the body?

4. How does the condition of the lens affect accurate vision?

5. Why can a deaf person have perfect balance?

6. How could a severe throat infection affect hearing and balance?

Appendix

Reference Tables: Metric System

1. *Linear Measure*

1000 meters	1 kilometer (km)
10 decimeters	1 meter (m)
10 centimeters	1 decimeter (dm)
10 millimeters	1 centimeter (cm)
1000 micrometers or microns	1 millimeter (mm)
1000 millimicrons or nanometers	1 micrometer (µm) or micron (µ)
10 Angstrom (A) units	1 millimicron (mµ) or nanometer (nm)
100 centimeters	1 meter
1000 millimeters	1 meter
2.54 centimeters	1 inch (in.)
1 millimeter	0.039 inches
1 meter	39.37 inches
1.6093 kilometers	1 mile

2. *Fluid Measure*

1,000 milliliters (ml.) or 1,000 cubic centimeters (cc.)	1 liter (l.)
1 liter	1.056 quart (qt.)

3. *Weight*

1,000 milligrams (mg.)	1 gram (gm.)
1,000 grams	1 kilogram (kg.)
453.6 grams	1 pound (lb.)
1 kilogram	2.204 pounds
28.35 gram	1 ounce (oz.)

4. *Temperature Conversion Scale*

To convert Centigrade (C.) to Fahrenheit (F.) multiply C. by 1.8 and add 32.
(Example: 37° C. to F.:)
F. = (37 × 1.8) + 32 = 98.6 F.
To convert Fahrenheit (F.) to Centigrade (C.) subtract 32 and divide by 1.8.
(Example: 68 F. to C.:)
C. = (68 − 32) ÷ 1.8 = 20 C.

Problems

1. With a cm. ruler measure the length of this zoology lab manual in centimeters. Convert this reading:

    – into millimeter(s) _____

    – into meter(s) _____

2. Convert 2.31 liters into milliliter(s) _____

    Convert 261 milliliters into liter(s) _____

3. Convert 866 grams into milligram(s): _____

    – into kilogram(s) _____

    – into pound(s) _____

4. Convert 2.50 pounds into kilogram(s): _____

    – into gram(s) _____

    – into milligram(s) _____

5. Convert 86° F. into C.

    Convert 100° F. into C. _____

    Convert 40° C. into F. _____

    Convert 100° C. into F. _____

Biological Terms

Learn the meaning and application of the following terms as soon as possible. These terms are used in describing the structures of animals and give direction for study and dissection.

aboral—pertaining to the region most distant to the mouth
anterior—the front, head end, or forward moving end of a bilaterally symmetrical animal
asymmetrical—an organism, body or body part that cannot be divided into two or more equivalent parts
bilateral symmetry—an organism, body or body part that can be divided into two equivalent parts, left and right
    halves, each a mirror image of the other
caudal—pertaining to or toward the posterior part or tail of an organism or body
cephalic—pertaining to or toward the anterior part or head of an organism or body
cross section—a section perpendicular to the long axis of the body
distal—away from the central part of the body or place of attachment
dorsal—pertaining to or toward the back or upper surface
lateral—pertaining to or toward one side
longitudinal—lengthwise or pertaining to the long axis of the body
median—pertaining to, near, or toward the middle line of the body
metamerism—body, externally, internally, or both, which is composed of a number of similar and homologous
    parts (segments, somites, or metameres) as seen in higher animal forms
oral—pertaining to the mouth
posterior—the hind part; toward the tail end of a bilateral symmetrical animal
proximal—toward or near the central part of the body
radial symmetry—an organism, body or body part having similar equivalent parts arranged around a common axis
sagittal plane—a plane that divides the body into symmetrical right and left halves
transverse plane—any plane at right angles to a sagittal plane; a cross section
ventral—pertaining to or toward the belly or lower side

Some Common Greek and Latin Prefixes, Suffixes, and Roots With Their Basic Meanings

a-; an-: not; without; an-emia; a-cephal-ous; a-poda

ad-; ac-; af-; ag-; ap-: to; toward; ad-hesion; af-ferent; ag-glutin-ate; ad-ductor; ap-pendage

ana-: up; ana-bol-ism; ana-phase; ana-tomy; an-oxia; an-ion

ant-; anti-: against; opposite; anti-biotic; anti-gen; anti-bodies

aqu-; aqua-; aquae-: water; liquid; aquat-ic; aqu-arium

-arium; -ary: place where something is kept, produced, or studied; api-ary; avi-ary; aqu-arium

arthr-; arthro-: joint (of body); arthr-itis; arthro-pod

auto-; aut-: self; spontaneous; auto-nomy; auto-nomic; auto-tomy; auto-troph

bi-; bin-: two; twice; double; bi-cuspid; bin-ocular

bio-: life; living; bio-logy; biot-ic; bio-genesis; bio-me; sym-bio-sis

blast-; blasto-: bud; cell; blast-ula; ecto-blast; osteo-blast; triplo-blastic

card-; cardi-: heart; cardi-ac; peri-card-ium

carn-; carno-; carni-: meat; flesh; carni-vore; carn-al

cephal-; cephalo-: head; cephalo-pod; a-cephal-ous; cephalo-thorax

chrom-; chromo-; chromat-: color; hue; chromo-some; chromat-id; chromat-in; chromo-nema

circum-: around; surrounding; circumpharyngeal

cyst-; cysti-; cysto: bladder; bag; sac; cyst; chole-cyst; sporo-cyst; tricho-cyst

cyt-; cyto-: cell; cavity; cyto-logy; cyto-kinesis; cyto-plasm; leuco-cyte; phago-cyte; choano-cyte

derm-; dermo-; dermat-: skin; hide; epi-dermal; derm-al; ecto-derm

di-; dis-; dipli-; diplo: two; twice; double; di-ptera; dipl-oid; di-morphic; dio-ecious; diplo-blastic; di-hybrid

di-; dia-; dis-: across; through; dissect; dia-phragm; di-gestion

e-; ef-; ex-: from; out of; without; ef-ferent; e-dentate; exo-skeleton; ef-fector; e-gest; ex-cretion

ec-; ecto-: from; outside; without; ecto-derm; ecto-zoa; ecto-plasm

end-; endo-; ent-; ento-: within; inside; endo-derm; endo-cardium; endo-crine; endo-genous; endo-skeleton; endo-thelium

ep-; epi-: on; upon; over; epi-cardium; epi-dermal; epi-glottis; epi-dermis; epi-thelium

-fer; -ferous: bearing; producing; ovi-ferous; omni-ferous; roti-fer; pori-fer-a

gam-; gamo-; -gamy: marriage; fusion; syn-gamy; iso-gamy; gam-ete; auto-gamy

gastro-; gast-: stomach; gastr-ic; gastro-pod; gastro-coel; gastro-dermis; gastr-ula

gen-: be born; producing; gen-es; genet-ics; anti-gen; geno-type; meta-gene-sis; oo-gene-sis; partheno-gene-sis

gloss; -glott: tongue; language; hypo-glossal; glott-is; epi-glottis

hem-; hemo-; hemat-: blood; hem-al; hemato-gram; hemo-rrhage; hemo-globin; hemo-coel

hepat-; hepato-: liver; hepat-ic; hepat-itis; hepa-rin

heter-; hetero-: other; different; hetero-genous; hetero-troph

homeo-; homo-: like; similar; same; homo-genous; homo-logue

hyper-: above; beyond; hyper-tonic; hypo-glossa; hyper-emia; hyper-dermic; hyper-trophy

hypo-: under; loss; below; hypo-tonic, hypo-glossal

is-; iso-: equal; similar; iso-tonic; iso-gamy

-log; -logy: word; discourse; study; bio-logy; zoo-logy; embryo-logy

mamm-; mamma-; mammae-: breast; nipple; mamm-al; mamm-ary

mes-; meso-: half; middle; mes-entery; meso-derm; mes-encephalon; meso-nephros

micro-; micr-: small; little; micro-scope; micro-be; micro-biology; micr-on

mon-; mono-: one; single; mono-saccharide; mono-cular; mono-ecious; mono-hybrid

nephr-; nephro-: kidney; nephr-idia; nephro-stome; nephr-itis; pro-nephros; nephr-on

nucle (nux): nut; nucle-us; nucleo-plasm; nucleo-tide

ocul-; oculus-oculi: eye; graft; ocul-ar; bin-ocul-ar; ocell-us; in-ocul-ation

-oid: like; in the form of; amoeb-oid; coll-oid; dipl-oid

omni-: all; omni-vorous

oss-; os-; ossi-; oste-; osteo-: bone; oss-eous; ossi-fy; osteo-blast; peri-oste-um; ossi-cle

ov-; ovum; ovi-: egg; ov-ary; ovi-duct; ovi-parous; ovo-viviparous

-parous: producing; ovi-parous; vivi-parous

pod-; podi-; podo-: foot; pseudo-pod; rhizo-pod; cephalo-pod; tetra-pod

poly-: many; much; poly-morph; poly-saccharide

pro-; proto-: first; pro-stomium; pro-tists; proto-plasm

pseud-; pseudo-: false; sham; pseudo-pod; pseudo-morph; pseudo-coel

sec-; sect-; seg-: cut; sec-tion; bi-sect; in-sect; vivi-sect; seg-ment

semi-: half; semi-permeable

soma-: body; chromo-some; auto-some

sub-: under; below; sub-esophageal; sub-lingual; sub-maxillary; sub-cutaneous

super-; sur-; supra-: over; beyond; more; supra-esophageal

ventr-; venter; ventri-: stomach; ventr-al; ventri-cle

vivi-: live; ovi-vivi-parous; vivi-sect

-vore; -vorous: eating; devouring; carni-vore; herbi-vore; omni-vore; insecti-vorous

zo-; zoa-; zoo-: animal; zoo-logy; proto-zoa; zoo; holo-zo-ic; metro-zo-an; spermato-zoa

zyg-; zygo-: pair; couple; yoke; zygo-te; zygo-spore; hetero-zyg-ous; homo-zyg-gous